D0519595
700044285861

the art of
sleeping

HQ

An imprint of HarperCollins*Publishers* Ltd.

1 London Bridge Street

London SE1 9GF

This edition 2019

1

First published in Great Britain by

HQ, an imprint of HarperCollins*Publishers* Ltd. 2019

A catalogue record for this book is

available from the British Library

ISBN: 978-0-00-833936-4

Printed and bound in Rotolito, Italy

Our policy is to use papers that are natural, renewable and recyclable products and made from wood grown in sustainable forests. The logging and manufacturing processes conform to the legal environmental regulations of the country of origin.

Rob Hobson

Registered Nutritionist Bsc, Msc

Book design by Steve Wells

An imprint of HarperCollinsPublishers Ltd.

TABLE OF CONTENTS

Dedicated to anyone
who has ever struggled
to sleep well

introduction

DREAMING OF SLEEP

Most of us spend one-third of our lives asleep, but not all of us sleep well. The amount of time we sleep and the quality of the sleep we get on a nightly basis can lead to tiredness and fatigue, the effects of which can filter into all aspects of daily life, affecting our emotions, our ability to focus on daily tasks, appetite, relationships and memory recall.

Many people underestimate the importance of sleep and live with the daily symptoms of fatigue, masking them rather than getting to the root of the problem. Thus, sleep deprivation has become an issue that's easily ignored, but if poor sleeping is left untreated it can have serious implications on diseases that can impact on your long-term health.

Sleep is the natural state of rest in which your eyes are closed, your muscles are relaxed, your nervous system is inactive and consciousness is practically suspended. This is a vital period of replenishment and repair for your body and a time when your brain is given the opportunity to process information, memories and experiences.

Sleep is essential, and it is undoubtedly one of the key pillars of good health. While our commitment to eating well and exercising regularly is ultimately down to personal choice and conscious decision-making, sleep is influenced by factors that are sometimes out of our control. You can make yourself a healthy lunch, you can find the drive to get yourself to the gym for an early-morning workout, but lying in bed and trying to fall asleep may be a little more tricky.

Many of us are victims of the 24-hour culture in which we live, as the modern-day demands and expectations of work and life in general, as well as the impact of social media, very much influence how we live our lives. This way of living can take its toll on our ability to sleep well and whilst you may think you are managing to survive on very little sleep, take it from me that you are not. Many of us have developed coping strategies to function on a daily basis (does that third cup of coffee before 11am sound familiar?) rather than taking a step back to address the real problem, which is not being able to sleep.

After years of struggling with insomnia, I became particularly interested in researching the different approaches to achieving the best possible sleep and one of the best pieces of advice I can give you is that one size never fits all.

In this personal, practical guide, which references the latest scientific research and expert opinion, I will break down the art of sleeping into three main pillars: Behaviour, Environment and Diet, which can be given the acronym, BED. Once you understand what your daily lifestyle looks like, it's possible to develop a sleep ritual that is personal to you.

My personal battle with sleep is what drove me to write *The Art of Sleeping*; however, I intend this book to be for everyone, whether you are similarly struggling with insomnia, looking for a better quality of sleep or are simply interested in the mechanics of sleeping well. I hope that by reading these pages you will be able to achieve the sleep of your dreams.

CHAPTER ONE

sleep

‘Sleep is that golden chain that ties health and our bodies together.’

Thomas Dekker

Sleep is a condition of body and mind that typically recurs for several hours every night, in which the nervous system is inactive, the eyes are closed, the postural muscles are relaxed, and consciousness is practically suspended.

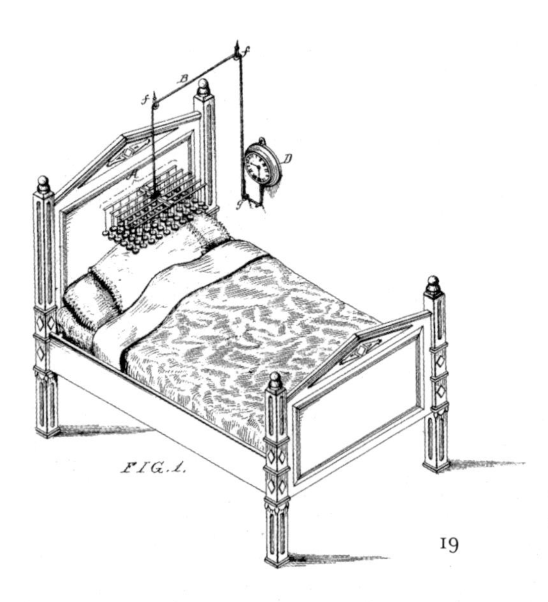
FIG. 1.

doze
slumber
siesta
nap
zizz

rest
snooze
bye-byes
drowse
kip
catnap

Sleeping is considered to be a time of rest, but your body is working hard to ensure you are kept in good health.

Your brain processes information, memory and experiences.

Growth hormone increases to help repair your body's tissues.

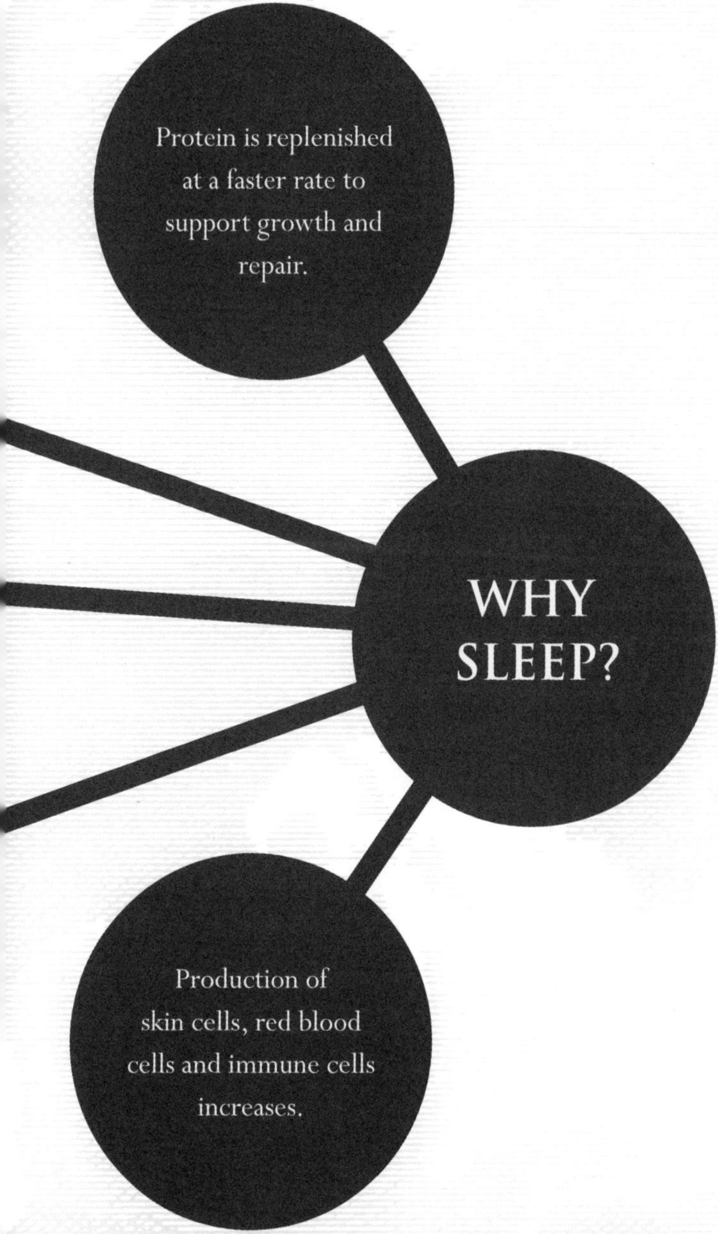

Protein is replenished at a faster rate to support growth and repair.

WHY SLEEP?

Production of skin cells, red blood cells and immune cells increases.

Sleep is essential to everyday life and influences many areas that impact on our day-to-day health and wellness, including:

Attention
Concentration
Creativity
Insight
Learning
Memory
Decisions
Emotions
Relationships

THE CIRCADIAN RHYTHM

Ever wondered why you feel sleepy at around the same time every night or wake up at the same time every day? It's simply part of your circadian rhythm at work.

Circadian rhythms are roughly 24-hour cycles that occur in the physiological processes of living beings – including plants, animals, fungi and cyanobacteria – and exist in every cell in the body, helping to set sleep patterns by governing the flow of hormones and other biological processes. Circadian rhythms are controlled by the body's internal clock and influenced by environmental factors such as light and temperature; the sleep/wake cycle is an example of a light-related circadian rhythm that determines our pattern of sleep.

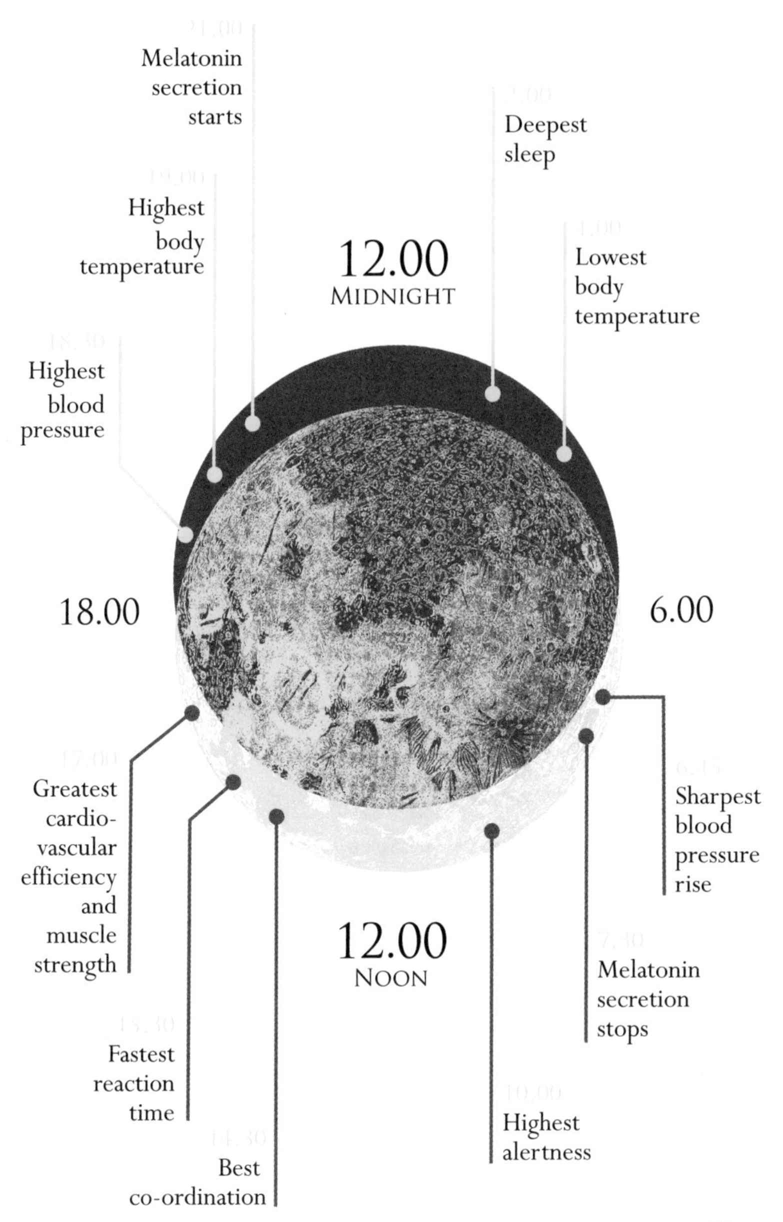

21.00
Melatonin secretion starts
19.00
Highest body temperature
18.30
Highest blood pressure
12.00
MIDNIGHT
2.00
Deepest sleep
4.00
Lowest body temperature
18.00
6.00
6.45
Sharpest blood pressure rise
17.00
Greatest cardio-vascular efficiency and muscle strength
12.00
NOON
7.30
Melatonin secretion stops
15.30
Fastest reaction time
10.00
Highest alertness
14.30
Best co-ordination

The modern human is thought to have originated just north of the equator in Africa, which is a region that has a constant 12 hours of daylight, and research has shown how evolution has impacted on our body clock. As humans migrated into a range of latitudes, they became exposed to variations in the length of daylight, which is thought to have influenced their biological clocks.

These rhythms are ingrained in us and make up the very fabric of our being. Wherever you live, the processes in your body are driven by the basic fact that every 24 hours the Earth pirouettes on its axis, creating a fixed pattern of sunlight and darkness. The knowledge that this clock keeps on ticking regardless of what's going on in our lives is quite comforting.

Under normal circumstances, the biggest energy dips happen in the middle of the night (somewhere between 2am and 4am) and just after lunchtime (around 1pm to 3pm), which is when many people crave a post-lunch nap. However, these times can vary slightly depending on your chronotype, which defines whether you're a morning lark or a night owl, which I will explain later (see page 59).

Sleep deprivation can make these fluctuations in sleepiness and alertness more noticeable, so if you're a good sleeper you're less likely to feel the dips as strongly as someone who doesn't get enough sleep.

Routine is essential in our day-to-day lives, as it helps us to keep in sync with the natural flow of our circadian rhythm. Going to sleep and waking up at the same time every day will keep your body in a steady state of flux, maintaining energy levels and ensuring proper regeneration throughout the body. Interrupted or erratic sleep will inevitably leave you feeling fatigued and out of sorts, while the effect of light can also influence your biological clock and circadian rhythm.

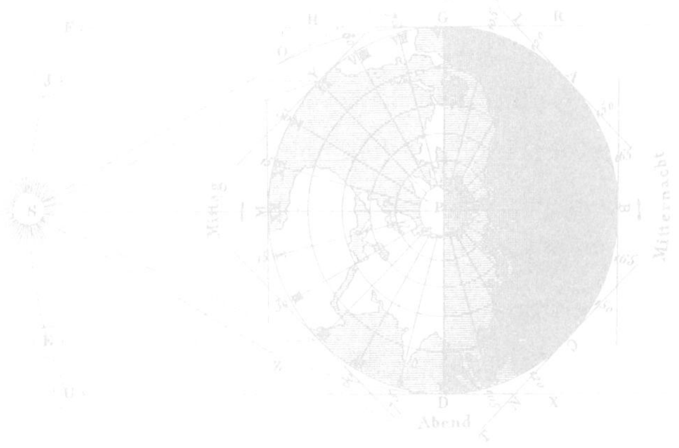

THE BODY CLOCK

Yes, this really is a thing! Your circadian rhythm can be thought of as a cycle of established events working in the background of your brain, but it's the complex action of nerve pathways in response to light that ensures they occur like clockwork.

Exposure to light stimulates a nerve pathway from the retina in the eye to an area in the brain called the hypothalamus. There, a special centre called the suprachiasmatic nucleus (SCN) works like a clock that sets off a regulated pattern of activities that affect the entire body such as the regulation of body temperature, heart rate, blood pressure and the release of hormones that help us to sleep.

MELATONIN: THE SLEEP HORMONE

Melatonin is a natural hormone made by your body's pineal gland and is the key hormone that drives our sleep/wake cycle.

This is a pea-sized gland located just above the middle of the brain; during the day the pineal is inactive, but when the sun goes down and darkness occurs, the pineal is 'turned on' by the SCN and begins to actively produce melatonin, which is released into the blood. Usually, this occurs between the hours of 9pm and 11pm. As a result, melatonin levels in the blood rise sharply at this time and you begin to feel less alert, making sleep more inviting.

Melatonin levels in the blood stay elevated for about twelve hours – all through the night – until the light of a new day breaks, when they fall back to low daytime levels by about 9am. Daytime levels of melatonin are barely detectable.

CORTISOL: THE WAKE HORMONE

Once exposed to the first light each day, the 'clock' in the SCN begins performing functions like raising body temperature and releasing stimulating hormones such as cortisol, made by your adrenal glands, which also encourage the uptake of the 'feel-good' hormone serotonin. The SCN delays the release of other hormones such as melatonin (which is associated with sleep onset) until many hours later when darkness arrives.

SLEEP ARCHITECTURE

The term 'sleep architecture' refers to the structural organisation of normal sleep. In the same way that your circadian rhythm can be characterized by a set of actions occurring in a cycle, so can the structure of your sleep, which occurs in different stages throughout the night.

Sleep can be divided into two groups: non-rapid eye movement sleep (NREM) and rapid eye movement sleep (REM). During NREM sleep your breathing and heart rate become slow and regular, your blood pressure drops and you remain relatively still.

As the name suggests, REM sleep is characterized by rapid eye movements as your pulse and breathing quickens, but the rest of your body remains motionless. It is during REM sleep that you're more likely to dream, and this is also the stage that occurs before you wake up.

A single sleep cycle is made up of four stages, each lasting around 90 minutes, which alternate cyclically throughout the night. The first three stages of the sleep cycle are NREM, each of which have their own set of unique characteristics, including brain wave patterns, eye movements and muscle tone – this takes up around 75 per cent of the cycle. REM sleep occurs in the fourth stage of the sleep cycle, taking up around 25 per cent.

STAGE ONE

Stage one is a short transition lasting only 5–10 minutes. During this unrestful stage, your eyes are closed but sleep is shallow, and you still have a sense of awareness. In stage one your brain is dipped into sleep, but you don't feel as though you are asleep. It is during this stage of sleep that you're most easily woken.

STAGE TWO

Stage two is often referred to as 'light sleep' and represents one of the most important parts of the sleep cycle, taking up almost half the night and characterized by a slowing down of both breathing and heart rate. Memories and emotions are processed during this stage, as is the regulation of your metabolism – the chemical processes that occur in the body to maintain life.

STAGE THREE

Breathing is slowest during this stage of NREM sleep and your muscles also start to relax, while heart rate is regulated. You're unlikely to be woken up in this stage, and if you are you will feel disorientated for a short while afterwards. The difficulty in waking up at this point in your sleep is one reason why your body tries to get deep sleep over with as quickly as possible. Your body has its own natural drive for deep sleep, so once you have met that, the need dissipates. The third stage normally occurs halfway through the night and your cycle then adjusts to more time in light sleep and REM.

These stages of sleep are very much about your body as the thinking parts of the brain go 'offline'. During deep sleep your body secretes the human growth hormone to help rebuild and repair cells of tissue, bone and muscle. Stages one to three also help to strengthen the immune system. Age can impact the sleep cycle, as you spend more time in light sleep and less time in these stages of deep sleep as you get older.

STAGE FOUR

While the previous stage of deep sleep is all about the body, stage four – or REM – is focused on the brain, because it is at this point in the sleep cycle that it is most active. Your body will largely remain inactive but your eyes will move rapidly in different directions. During this stage your heart rate increases and your breathing becomes more irregular. Protein synthesis also peaks, helping to maintain the processes required to keep your body working properly. Dreaming usually occurs in this fourth stage, as well as the regulation of emotions and memories.

DREAMING

Dreaming is one of the most notable but least understood characteristics of sleep, during which our thoughts follow bizarre and seemingly illogical sequences, sometimes random and sometimes related to experiences gathered during wakefulness. While the most intense dreams occur during REM sleep, because this is when the brain is most active, some can still occur during the stages of NREM.

Dreams often take on a fantastical feel, as within them we're able to act out scenarios that would never be possible in real life. However, the experience is not always positive, and nightmares can induce feelings of terror, anxiety and distress, which have been linked to sleep problems such as insomnia.

There are many explanations as to why we dream, and these have been offered by both philosophers and psychologists. Sigmund Freud suggested that dreams reveal a person's deepest unconscious desires and that we disguise these impulses with symbolic objects. Other theories offered by researchers have suggested that dreams may be a type of offline memory processing, whereby the brain consolidates learning and daily memories, and that dreams even offer a way of developing cognitive capabilities. It has also been suggested that dreams are an ancient biological defence mechanism, simulating threatening events so that we're more perceptive and able to avoid them in real life. However, others believe that dreams are simply a result of random activity in the brain.

The true meaning of dreaming is still something of a mystery and many questions remain unanswered by the current available research. Perhaps we will never know, but for now you can choose what you would like to believe.

B

THE STUFF OF NIGHTMARES

Most people don't get enough sleep. We are a society that burns the candle at both ends, a nation where people stay up all night to study, work or have fun.

However, going without adequate sleep per night carries with it both short- and long-term consequences that can affect every aspect of our lives.

The optimum number of hours of sleep is thought to be just under eight, but research carried out by the Royal Society for Public Health has shown that most people manage less than seven. Over the course of a week this deficit equates to a whole night's sleep, and research by The Sleep Council has shown that 33 per cent of people only manage 5–6 hours, while 7 per cent get less than 5 hours.

The regenerative power of sleep allows the brain to process information, muscle and joints to recover and enables protein to be replenished in every part of the body, which promotes the growth and repair of tissues, cells and organs. Most of us are familiar with the short-term effects of not getting enough sleep, when we experience fluctuations in mood, concentration and alertness, and our ability to recall memories, our creativity and decision-making are also affected. All of these can filter through into many areas of everyday life, such as relationships and work. But it's the long-term effects of poor sleep that will really give you nightmares.

DIABETES

Research published in the journal *Sleep Medicine Clinics* found that insufficient sleep could put you at a greater risk of type 2 diabetes by affecting the way your body uses glucose, the carbohydrate fuel that energizes cells. The study showed that when healthy subjects had their sleep cut in half from eight to four hours per night, they processed glucose more slowly than when they slept for twelve hours. It's a finding that is reflected in many other studies of a similar nature.

HIGH BLOOD PRESSURE

Inadequate sleep can also cause elevated blood pressure, even when this happens over minor periods of time. A study carried out by the University of Alabama found that a single night of inadequate sleep in people who already have high blood pressure can cause elevated levels the following day. As high blood pressure is a risk factor for heart disease and stroke, this contributes to the correlation between poor sleep and heart disease.

MENTAL HEALTH

Your mental health can also be affected by lack of sleep. Given the effect a sleepless night can have on your mood and concentration, it's not too much of a leap to think that chronic sleep deprivation may result in more serious mood disorders. Already there is lots of well-documented research showing an association between chronic sleep issues and depression, anxiety and mental distress. In one study carried out by University College London, subjects who slept only four hours per night showed declining levels of optimism and sociability following repeated days of inadequate sleep. In a similar study, subjects with less than four hours' sleep reported feeling sadder, stressed, angry and mentally exhausted. Notably all of these symptoms improved dramatically when they returned to a normal pattern of sleep.

WEIGHT GAIN

If you're putting on weight or finding it difficult to lose weight, research suggests this may have something to do with a lack of good-quality sleep. Studies carried out by Loughborough University found that people who habitually sleep for less than six hours per night were more likely to have a higher than average Body Mass Index (BMI) while those who got eight hours had the lowest BMI score.

It's becoming more widely accepted that along with a lack of exercise and poor diet, a lack of sleep may potentially be just as influential in the development of obesity. This is because lack of sleep is thought to impact on the hormones leptin and ghrelin which control our appetite and may play a role in weight gain. Leptin, often referred to as the satiety hormone, is released from fat cells and sends signals to the hypothalamus in the brain, which helps to inhibit hunger and regulate energy balance so the body does not trigger hunger responses when energy is no longer needed. Ghrelin is often referred to as the hunger hormone and is released by the stomach to stimulate appetite, thus increasing food intake and encouraging fat storage. Research has suggested that a lack of sleep reduces leptin and elevates ghrelin, which may explain the correlation between obesity and a lack of sleep highlighted by certain studies.

A lack of sleep may also impact on the release of other hormones related to weight gain, such as insulin and cortisol. Insulin regulates blood glucose (sugar) but also promotes the storage of fat and, as such, higher levels of insulin have been associated with weight gain and the risk of developing diabetes. Research has shown how a lack of sleep may increase the secretion of insulin after you eat a meal, and also cortisol (the stress hormone), which has long been associated with encouraging the body to store fat.

A lack of sleep may increase your energy intake by 300 calories per day.

Of course, a lack of sleep also causes tiredness and fatigue, which may hamper the motivation required by some people to exercise or eat healthily, both of which can impact on body weight.

MIND OVER MATTER?

A continuous lack of sleep can start to become a bit of a mind game over time, and it can also cause unwanted stress and anxiety that can be hard to deal with.

Catch yourself if phrases such as 'I'm so tired', 'I hardly slept last night' or 'I was awake for hours' start to become a part of your daily dialogue. If you're not sleeping properly you need to take matters into your own hands and change things rather than letting it define you and relying on ways of adapting to daily life without sleep.

A chronic lack of sleep is defined by the number of hours you manage to get each night. The definition of sleep loss is defined by getting anything less than the recommended eight hours. Emotional distress can make matters worse and it becomes a vicious cycle as the worry or anxiety about not sleeping makes it even more difficult to fall asleep.

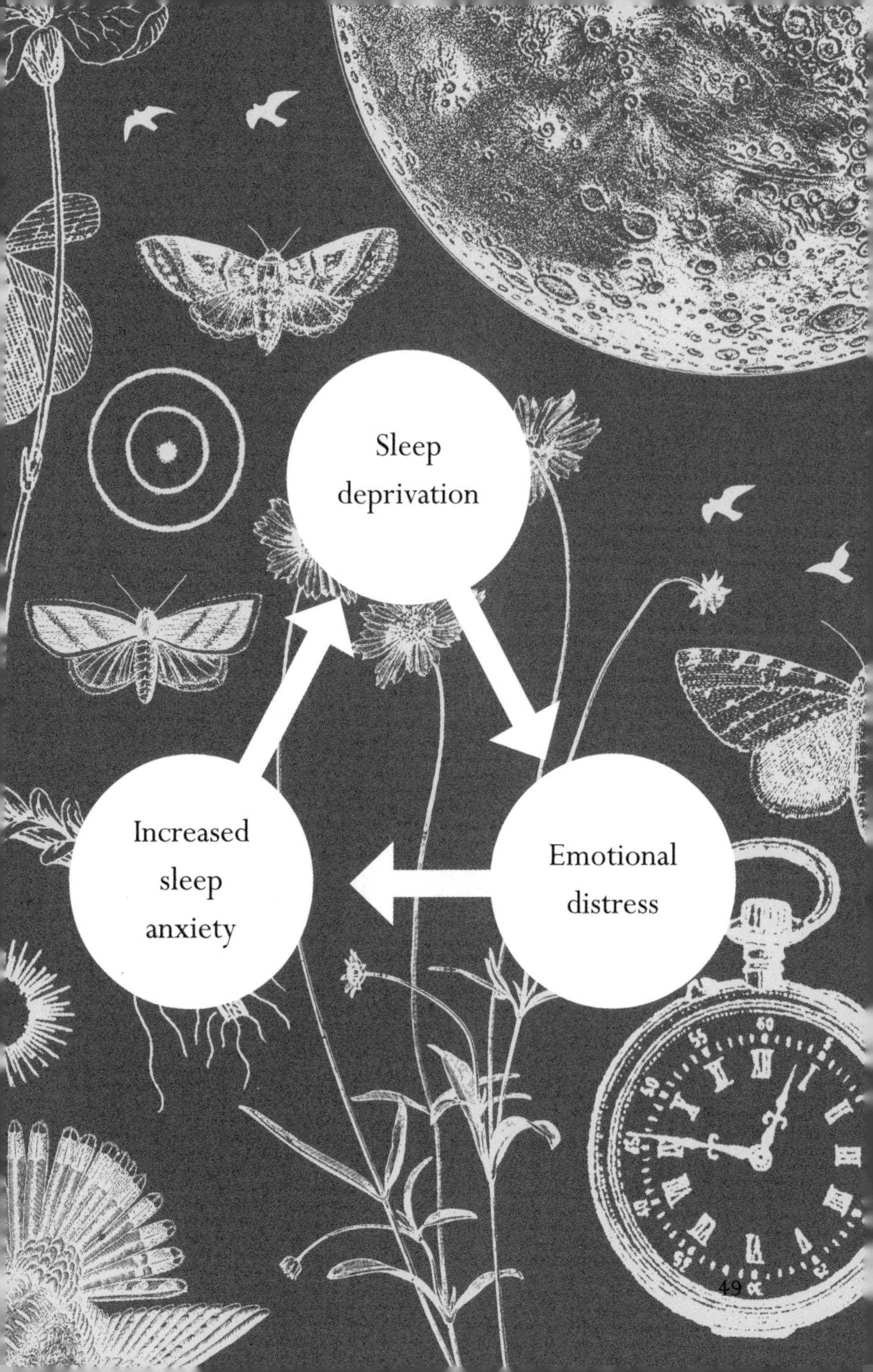
Sleep deprivation
Emotional distress
Increased sleep anxiety

People who suffer with insomnia often hold unrealistic expectations about their sleep requirements and worry excessively when these are not met.

The most important thing you can do is break the cycle and do something about it!

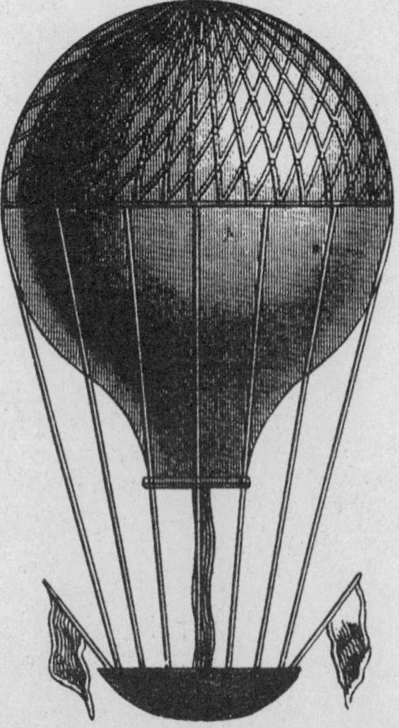

AT NIGHT

Rather than lying in bed counting sheep, get up and distract your mind from the anxiety of not sleeping. Find a quiet space and limit your exposure to bright light. Make yourself a warm drink or whatever works best to help you relax. Later in the book I talk about jotting down your thoughts or anxieties, which can help to clear the mind. Other gentle activities, such as reading, can also be a useful way to distract the mind and instigate the tiredness required to get you back to sleep.

IN THE DAY

It's far too easy to blame everything on your lack of sleep, but this isn't helpful. If you're worrying about your sleeping habits during the day, of course it's going to be even more difficult to fall asleep at night. Instead, focus on the challenges ahead rather than how tired you're feeling. Stay positive, change your language and keep perspective.

Sometimes if your sleeping habits are becoming unmanageable, it's important to adapt your lifestyle to find a routine that works for you. And that's where the art of napping may be a useful way of tackling daily fatigue when it hits, whilst also working with your innate pattern of sleep.

CHAPTER TWO

napping

‘You must sleep some time between lunch and dinner, and no halfway measures. Take off your clothes and get into bed. That’s what I always do.’

Winston Churchill

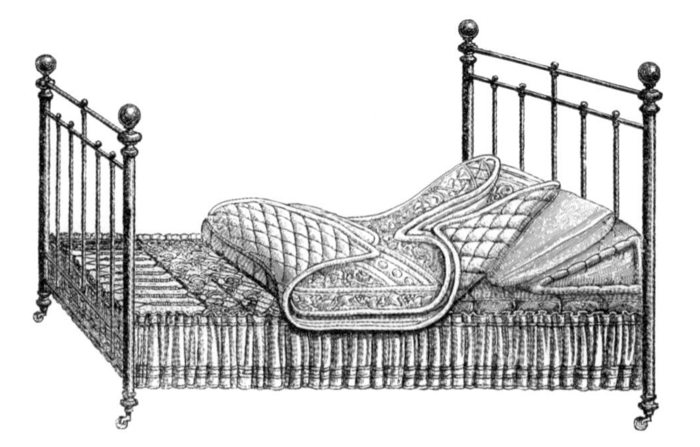

THE ART OF NAPPING

The majority of mammals are polyphasic sleepers, sleeping multiple times within 24 hours. Humans, on the other hand, are monophasic sleepers, sleeping only once during the same time period.

However, it is still unclear whether this is our natural sleep pattern, especially if you look at the sleeping habits of young children and the elderly, who take regular naps throughout the day. In fact, research suggests that our ancestors were polyphasic, sleeping for shorter periods, but over time we have adapted this way of sleeping into a single block to complement and cope with the frenetic pace of our modern existence.

Napping is still an important part of cultures in the Mediterranean, South America and Africa, where many people traditionally adopt the polyphasic approach to sleep.

The benefits of napping include restoring alertness, enhancing performance and overcoming fatigue, while psychologically it can be seen as a luxury enjoyed during time off to help rejuvenate and relax. Historically, the afternoon nap has been favoured by many intellects and leaders, such as Winston Churchill, Napoleon, Einstein and John F. Kennedy.

FORTY WINKS

Napping for up to 30 minutes during the day can help you to reap the benefits if you need to do so, and in line with the concept of planning your sleep in cycles (explained later on page 69), this can help you to catch up on sleep lost for every 90 minute cycle missed during the night.

The best time to slot a nap into your day is between 1pm and 3pm to follow the natural cycle of your circadian rhythm (remember from the circadian rhythm diagram on page 27 that this is when the body is naturally more relaxed and sleepier).

THE POWER NAP

The power nap is something advocated by many sleep experts. It is beneficial for a quick jolt of alertness and is a useful way to ward off those moments of fatigue that make it almost impossible to keep your eyes open. In some cases, simply shutting your eyes for 10 minutes is all you need to counter the effects when fatigue hits you in this way.

YOU SNOOZE, YOU LOSE

Napping isn't for everyone, and snoozing for any longer than 30 minutes can mean running the risk of entering into a deep sleep. The feeling of grogginess and disorientation associated with napping for too long is often referred to as sleep inertia and can last for up to 30 minutes, which isn't great if you have something important to get on with. If you have trouble sleeping, napping for long periods of time during the day may also adversely affect the length and quality of your sleep at night.

LARK OR OWL?

Research has suggested that our sleep patterns are actually programmed by our DNA.

Our propensity to sleep at a particular time during a 24-hour period has been defined by scientists as our chronotype and is specifically linked to the PER3 gene. Two common chronotypes used to define the way people sleep are referred to as being a 'lark' or an 'owl'. Morning larks have longer PER3 genes, needing more sleep, whilst night owls have shorter PER3 genes, needing less.

Research involving thousands of participants has also shown that larks and owls may share personality traits with one another. While larks may be more punctual or conscientious – often more of a 'go-getter' – the owl's tendency for openness and risk-taking can be positively correlated with creativity.

Thousands of years ago, our ancestors would have benefitted from the different sleep chronotypes. Lark and owl tendencies would have meant that there was always someone awake to keep watch for danger. Fascinating, isn't it?

- Goes to sleep 9pm–11pm
- Wakes 5am–7am
- Natural wake
- Loves mornings and breakfast
- Less daytime fatigue
- Tend to be more conscientious, cooperative and persistent
- A go-getter
- Probably not a procrastinator

NIGHT OWL

- Goes to sleep 12pm–3am
- Wakes 9am–11am
- Alarm wake
- Loves evenings and dinner
- Daytime nappers
- Seeks out novelty and originality
- Tends to be more open to creative thinking
- Risk-taker and prone to addictive personality
- Tendency to procrastination

These characterisations of sleep chronotypes represent a simple polarization between people who are more active during the day (morning larks) and those more active at night (night owls). Research carried out by the psychologist Dr Michael Brues on the topic of sleep chronotypes has delved deeper to define four other descriptive categories classified as the Dolphin, Lion, Bear and Wolf, which sit somewhere else on the spectrum.

DOLPHIN

thought to represent around 10% of the population

These are light sleepers with a low sleep drive who struggle with sleeping through the night, waking up repeatedly, and have anxiety-related insomnia. This group mull over mistakes or things they have said over the day as they lie awake at night. This group works best alone, and dolphins are averse to confrontation. Dolphins tend to have a low to average body weight and care less about fitness as a result.

- Cautious, introverted, neurotic, bright
- Avoid risk and strive for perfection and detail
- Wake up unrefreshed and continue to feel tired until late in the evening
- Most alert late at night, with productive spurts throughout the day

LION

thought to represent around 15–20% of the population

Like one of nature's apex predators, lions rise before dawn, starving, and after a hearty breakfast are ready to conquer the goals they have set for the day ahead. Lions are focused, with a clear strategy and objectives to face challenges head-on to achieve success. Good examples of lions include entrepreneurs or CEOs. Exercise is important to this group and fits into their ethos of achieving goals.

- Conscientious, practical, stable, optimistic
- High-achieving and interactive, they make health and fitness a priority
- Early risers who suffer with mid-afternoon energy slumps and fall asleep easily
- Most productive in the morning and most alert at midday

BEAR

thought to represent around 50% of the population

Outside of hibernation, bears are active in the day and restful at night, which is known as diurnal. Like the bear, this group seek to sleep for at least eight hours every night, if not more. Feeling fully awake can take a few hours, during which they feel hungry. Their appetite reflects that of the bear, as they feel hungry most of the time and, if food is available, will eat regardless of whether it is a mealtime. This group are friendly, good-natured and easy to talk to, making them perfect party guests and the least likely to cause a fuss at work or blame others for their mistakes.

- Cautious, extrovert, friendly, open-minded
- Avoid conflict, aspire to be healthy, prioritise happiness and like familiarity
- Wake up dazed and usually hit the snooze button
- Tiredness hits mid-to-late evening; they sleep deeply but always crave more rest
- Most alert between mid-morning and early afternoon
- Productivity peaks just before midday

WOLF

thought to represent 15–20% of the population

Like wolves in nature, this group comes alive when the sun goes down. While unlikely to be hungry on waking, wolves are usually ravenous during the night. This group is the most likely to make poor food choices, and along with their preferred eating schedules are more likely to be overweight while also at greater risk of diet-related diseases. The wolf is creative but unpredictable, and is easily insulted by the perception of being lazy, as well as being more susceptible to depression and anxiety as a result of their nocturnal lifestyle.

- Impulsive, pessimistic, creative, moody
- Risk-taking, pleasure-seeking and reacts with high emotion
- Struggles to wake before midday and doesn't feel tired before midnight
- Most alert after 7pm
- Most productive late morning and late evening

If a person's tendency to one particular chronotype is very strong and seriously impacts on their daily routine, then they may be diagnosed with a circadian rhythm sleep disorder (CRSD). This isn't insomnia as they may sleep perfectly well, only that they do so at a delayed time as their body clock is substantially off from the rest of society. People with a CRSD can be forced into waking up earlier than the body is prepared for or wake up very early, which can cause fatigue during the day and interfere with work and social demands.

Understanding your chronotype may help you to organize your daily life to work in the most beneficial way for you. Rather than work against your body's natural tendencies towards sleep by forcing yourself to stay awake or go to bed too early, try and schedule meetings or social events around the times at which you're most productive and energized (although in reality this may not be so simple to navigate).

It's important to keep in mind that whilst these definitions of sleep chronotypes may help to explain common traits amongst the population, they don't reflect upon your values as a person and certainly don't predict success in relation to sleep patterns.

Fig. 34
F
a
b
c
d
e
f
g

SLEEP CYCLES

We all know that we should be aiming for around eight hours of sleep a night in order to allow the body to fully rest and regenerate. However, if you have difficulty sleeping, the pressure to meet this quota can, as we have discussed, play a key part in the vicious cycle that is keeping us awake.

When it comes to sleep, it's about quality, not just quantity. Waking up refreshed, alert and energized requires the right type of sleep, too.

Rather than sticking to the 'eight-hour rule', another approach is to structure your sleep into 90-minute cycles.

The first step is to establish a routine which will work in synchronicity with your circadian rhythm, to calculate the number of sleep cycles you need to achieve across the week to meet your sleep goals.

In order for this to work, you need to commit to a set wake-up time then work backwards to establish the time at which you need to go to sleep. For example, if your goal is five sleep cycles (7.5 hours' sleep in total) and you aim to wake up at 6.30am, you should be looking to fall asleep at 11pm. A sleep starting point between 9.30pm and 11pm works in sync with the natural flow of your circadian rhythm, as this is when the body starts to lower serotonin levels and increases melatonin to induce drowsiness.

DISCIPLINE IS KEY

Training yourself to find a set rhythm of sleep is the crux of the 'Sleep Cycles' method, and that means going to bed at the same time every night and waking at the same time every morning. Very soon you will notice how your body naturally wakes up at the same time every day.

MORE ISN'T ALWAYS BETTER

Often it's tempting to try to 'catch up' on sleep at the weekend, but this can actually leave you feeling worse.

As we now know, it's hugely important to wake up at the same time every day, to avoid disrupting your body clock.

Sticking to a regular wake-up time means your body can wake itself up without the need for an alarm clock. In the hour before you wake, your sleep becomes lighter, your body temperature rises and cortisol levels start to rise, too, providing you with the energy you require to wake up.

AVOID HITTING SNOOZE

As much as you can, try to resist hitting the snooze button after your alarm goes off! Falling asleep again will leave you feeling groggy because you put your body and brain out of sync with their natural rhythm.

Of course, for non-sleepers, the issue is not *when* but *how* to fall asleep...

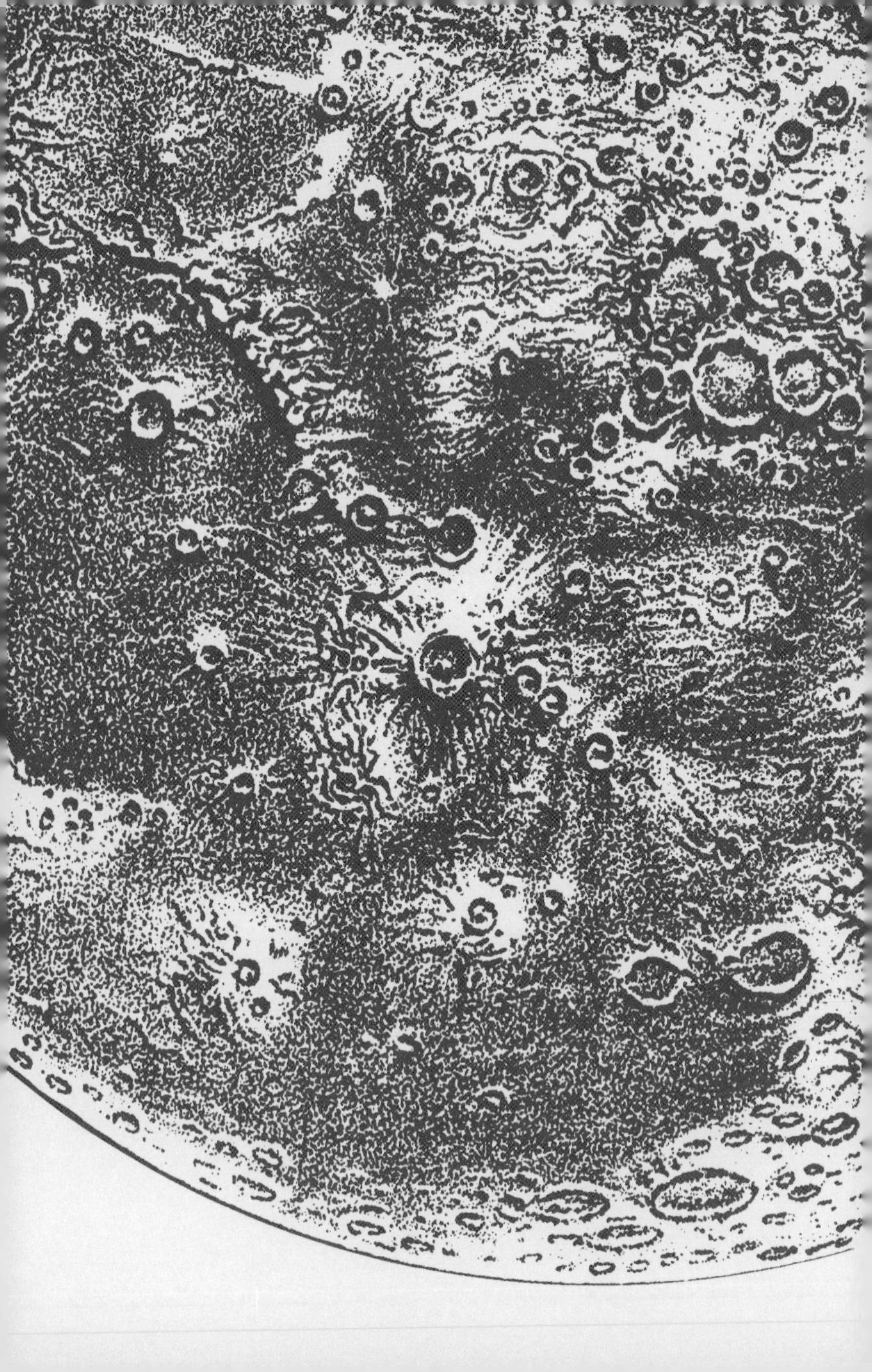

CHAPTER THREE

bedtime

DEAR DIARY

If you have trouble nodding off or staying asleep when it is time for bed, it is important to first identify what might be getting in your way.

Whether it is your BEHAVIOUR, the ENVIRONMENT or your DIET, completing a personal sleep diary is the most effective way of capturing all of these contributing factors so that you can start to think about your unique sleep ritual. Over seven days, complete two diaries: one when you wake up and the other before you go to sleep.

The Morning Diary will help you establish your usual bedtime and the number of hours you slept, as well as the frequency with which you woke during the night (as well as any possible reasons for this). The Evening Diary will help you to identify any lifestyle factors that may be getting in your way of a good night's sleep.

Once you have a better understanding about what it is that is getting in the way of a good night's sleep you can use the information in the following chapters to develop new habits that form the basis of your unique sleep ritual.

FILL IN EVERY MORNING	EXAMPLE	MONDAY	TUESDAY
BEDTIME	10pm		
MORNING TIME OUT OF BED	7.30am		
ABILITY TO FALL ASLEEP:			
• Easy			
• Took some time	X		
• Difficult			
AWAKE DURING THE NIGHT			
• Number of times	2		
• Total time awake	3 hours		
• Total sleep (hours)	6.5 hours		
REASONS FOR DISTURBED SLEEP *(list all physical and mental factors, such as noise, stress, light, partner snoring, discomfort, temperature, physical pain – joints, digestion)*	Noise (partner snoring), worry		
HOW DO YOU FEEL ONCE AWAKE?			
• Fully refreshed and energized			
• Moderately refreshed	X		
• Fatigued			
LIFESTYLE FACTORS *(list any other reasons that are affecting your sleep, such as work hours, monthly cycle, security worries, busy mind)*	Financial worries made it difficult to nod off and fall back to sleep once awake during the night.		

WEDNESDAY	THURSDAY	FRIDAY	SATURDAY	SUNDAY

FILL IN EVERY EVENING BEFORE BED	EXAMPLE	MONDAY	TUESDAY
NUMBER OF CAFFEINE DRINKS			
• before 5pm	3		
NUMBER OF CAFFEINE DRINKS			
• after 5pm	1		
ALCOHOLIC DRINKS AFTER 5PM: *1 unit is equivalent to 1/2 pint beer, single shot of spirit, 1/2 small glass of wine (76ml) or 250ml of alcopop*			
• 1–2 units	X		
• 3–4 units			
• More than 4			
ANY MEDICATIONS TAKEN DURING THE DAY *and what were they?*	None		
NAPS DURING THE DAY *(answer yes or no and the length of time)*	Yes, 1 (30 minutes)		
DID YOU FEEL ANY OF THE FOLLOWING DURING THE DAY?			
• Tired			
• Moody	X		
• Impatient			
• Unable to focus or concentrate			
BRIEFLY DESCRIBE YOUR ROUTINE IN THE HOUR BEFORE BED	Had a bath just before bed then watched a box set in bed on my laptop for a couple of hours. Checked my work emails just before switching the lights off.		

WEDNESDAY	THURSDAY	FRIDAY	SATURDAY	SUNDAY

behaviour

environment
diet

CHAPTER FOUR

behaviour

‘Think in
the morning.
Act in the noon.
Eat in the
evening.
Sleep in
the night.’

William Blake

Your evening routine
before you go to bed can
influence the quality and
duration of your sleep.

Your morning routine after
waking can have an impact
on the rest of your day.

Think of your bedroom
as a 'slumber palace',
reserved exclusively for
sleep. And yes, sex has
been shown to promote
a good night's sleep.

d
a
b
WINTER. 95 Tage.
FRÜHLING. 90 Tage
Mittlere Temperatur
Januar
Februar
März
April
Widder
Stier
Fische
Wassermann
Fig. 1

LIGHTS OUT!

In 1981, a Harvard Medical School professor, Dr Charles Czeisler, discovered that it is daylight that keeps our circadian rhythm, or body clock, aligned with our surroundings.

Any light can suppress the secretion of melatonin so try and keep your bedroom dark by using blackout blinds or investing in a sleep mask. If you wake up during the night, then any light creeping through gaps in curtains and blinds can be a distraction preventing you from getting back to sleep.

While any light can suppress the secretion of melatonin – the hormone that promotes sleepiness – it is blue light that has the greatest negative effect. This light is omitted from electrical equipment such as computers, mobile phones, notebooks and TVs.

If you do need a light on, then research has shown how red light has the least impact on melatonin production making this wavelength of light the most conducive to sleep.

You can buy red or pink bulbs to use in your bedroom and even strings of novelty lights, but these may not be to everyone's taste. The next best thing is to use incandescent bulbs that give off a diffused, warm light and can be controlled with a dimmer switch on side lamps.

During the day, make sure you expose yourself to plenty of natural light as this can help to boost mood and make you feel more energized. This in turn can have a positive effect on your ability to sleep at night.

But remember, when it's time to sleep, it's lights out!

DIGITAL DETOX

Of course, for many of us it's not the main bedroom light that's the problem, it's the phones, laptops and televisions we're using before we go to bed.

Settling in to watch a box set, following the news, catching up on emails or scrolling through your preferred social media are just a few examples of how digital technology has become an intrinsic part of modern life. But the downside is that they might be playing a harmful role in our ability to sleep well.

As long as you're tuned in to your smartphone or laptop, you're essentially making yourself permanently available, and this can make it difficult to switch off and put your mind at rest. Do you really need to check your emails last thing at night? Do you really need to see a picture of what someone ate for dinner just before bed? Can it really not wait? Recent studies show that we check our phones anything between 80 to 200 times a day. A survey of over 4,000 British adults carried out by Deloitte in 2017 found that 38 per cent

thought they were using their
phone too much and amongst
those aged between 16 and 24 years
old this figure rose to more than
half. As many as 79 per cent of adults
said they checked the apps on their phone in
the hour before they went to sleep and 55 per cent
did so within 15 minutes after waking up.

OUT LIKE A LIGHT

A study published in the *Proceedings of the National Academy of Sciences of the United States of America* showed that short-wave (blue) light emitted by tech devices shifted the phases of the circadian rhythm and suppressed melatonin. This increased alertness before bed, affected the time it took for people to fall asleep and shortened the REM stage of their sleep cycle.

The study also found that even after getting eight hours' sleep, people exposed to more blue light before bed were sleepier and took longer to wake up. People who use electronic devices before bed have also been shown to stay up later, affecting both their circadian rhythm and sleep time.

FEELING BLUE

Research has identified a correlation between the overuse of smartphones and depression, especially in younger adults. The over-reliance and, in some cases, addiction to the use of smartphones has also been shown to influence other areas affecting mental health, such as anxiety, obsessive compulsive behaviour and interpersonal sensitivity. Every one of these effects can impact your ability to sleep.

Of course, it's completely unrealistic to extinguish tech devices from our lifestyle, but finding strategies to manage how we use them can help you in your path to good sleep. There is little doubt that on completion of your sleep diary the use of tech devices will come up as an issue. As part of your sleep ritual, instigate a personal curfew, one that can be widened to the whole household if necessary. Switch your device to 'night mode' and set the time restriction for no 'blue light' devices two hours before bed, as this will give you the best chance to eliminate this as a factor in your sleep problem.

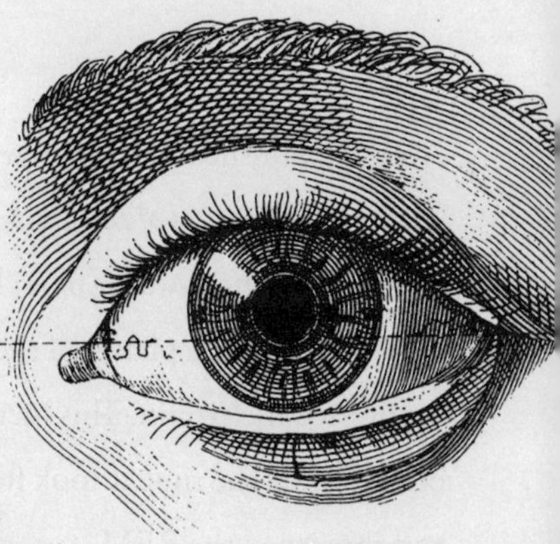

RISE AND SHINE

It's not just about getting to sleep, don't be in a rush to plug yourself back into the digital world when you wake up. You should do everything you can to start the day in the natural flow of your circadian rhythm. When you wake up, open the curtains to allow as much natural light to enter your bedroom as possible, which will encourage the brain to stop releasing melatonin as it uses the hormone cortisol as a tool to help us to rise from slumber and encourage appetite.

Try not to switch your phone on until you have showered and eaten breakfast, because starting your day on a sour note as a result of a negative email just encourages stress. The effect of stress then causes the body to over-produce cortisol, which can put your circadian rhythm out of sync for the day ahead while also impacting on mood and, in some cases, appetite.

CHILL OUT

If you want to prepare your body for a good night's sleep, you need to chill out. When we think about the effect of temperature on our body it's easy to assume that heat can help us to sleep. Sitting outside in the midday sun or inside a hot stuffy office can leave you feeling dozy, but if you're trying to fall asleep in the evening, heat can make things difficult.

The tiredness you feel from high external temperatures during the day is a side-effect of your body trying to cool you down. Your body responds to these high temperatures by expanding blood vessels, which increases blood flow nearer to the skin to release heat and cool the body. At the same time, your blood pressure drops, resulting in less oxygen being delivered to various systems in the body, which causes fatigue.

In contrast, your circadian rhythm is very attuned to body temperature – it's one of the functions it controls to help you fall asleep or stay awake. During the day, your body temperature rises naturally until late afternoon, at which point it then starts to fall. As you start to fall asleep your body temperature begins to lower by one to two degrees, which helps the body to conserve energy. This drop in temperature signals the release of melatonin to help induce relaxation and sleep by slowing the heart rate, breathing and digestion. If your sleep environment is too hot or cold, this can make it more difficult for your body to reach the optimal temperature required for a good quality of sleep.

TAKE A BATH

While it may seem counterintuitive to what we've just discussed, many studies have shown that warming your body by bathing can help to promote sleep, but to harness these effects, timing is key. The best time to take a bath is at least one hour before you hit the hay, as this gives your body enough time to cool down to its optimum sleep temperature. Similar effects have been shown when showering or even soaking your feet in warm water to increase your skin and body temperature.

Bathing has also been shown to help relieve anxiety and muscle stress, which can help with relaxation and sleep. Epsom salts are a good choice for putting in the bath water, as they are rich in magnesium which helps to promote muscle relaxation and sleep.

Bath oils can also help with relaxation as they stimulate the olfactory nerve. This nerve gives us our sense of smell and sends signals to parts of the brain that are in charge of emotions and mood, soothing us through the parasympathetic nervous system which relaxes the body. Oils traditionally used for relaxation include lavender, bergamot, ylang ylang, clary sage and vetivert. Whilst the essential oil aromatherapies may not have been rigorously studied, they can still have calming effects.

You can make bath time even more relaxing by burning candles and turning off the bathroom light. Listening to calming music or using a meditative app on your phone can also make bath time even more relaxing and offer an opportunity to calm a busy mind.

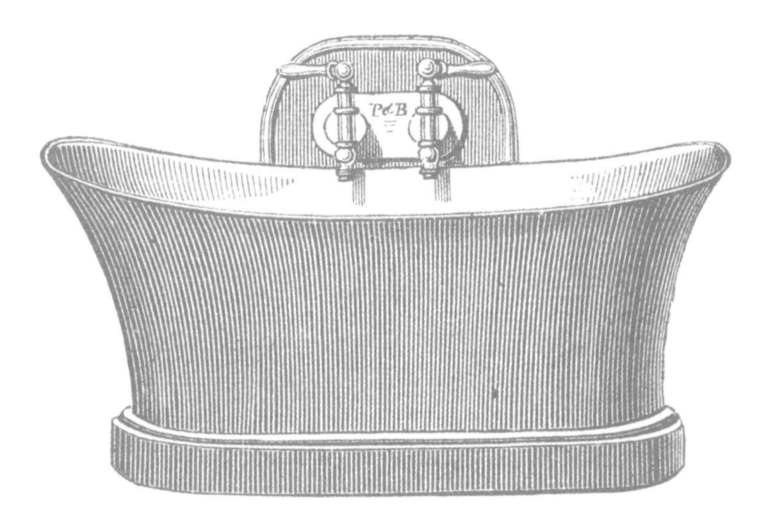

BRAIN DOWNLOAD

Restlessness and a busy mind can easily make falling asleep difficult. As you lie awake your mind can go into overdrive while you focus on the issues and worries impacting on your life, many of which you will unconsciously ruminate on all night.

People who write down their thoughts, activities and tasks that need to be completed before they go to bed fall asleep much quicker than those who don't.

Keep a pad of paper and a pen next to your bed so you can jot down your thoughts before you go to sleep each night. As well as writing down your worries and stresses, include any unfinished tasks that need to be completed the following day, or make a to-do list.

If you wake up during the night and your mind starts to wander, read through your diary and to-do list, adding to it if you need to. Sometimes the best ideas can occur in the middle of the night, so be sure to keep plenty of space to jot these down.

As I mentioned earlier, don't spend hours lying in bed trying to fall asleep. Instead, get up and sit somewhere quiet, keeping the lights down low. Use this time to reflect and to help organize your thoughts by writing them down rather than letting them buzz around on repeat in your head.

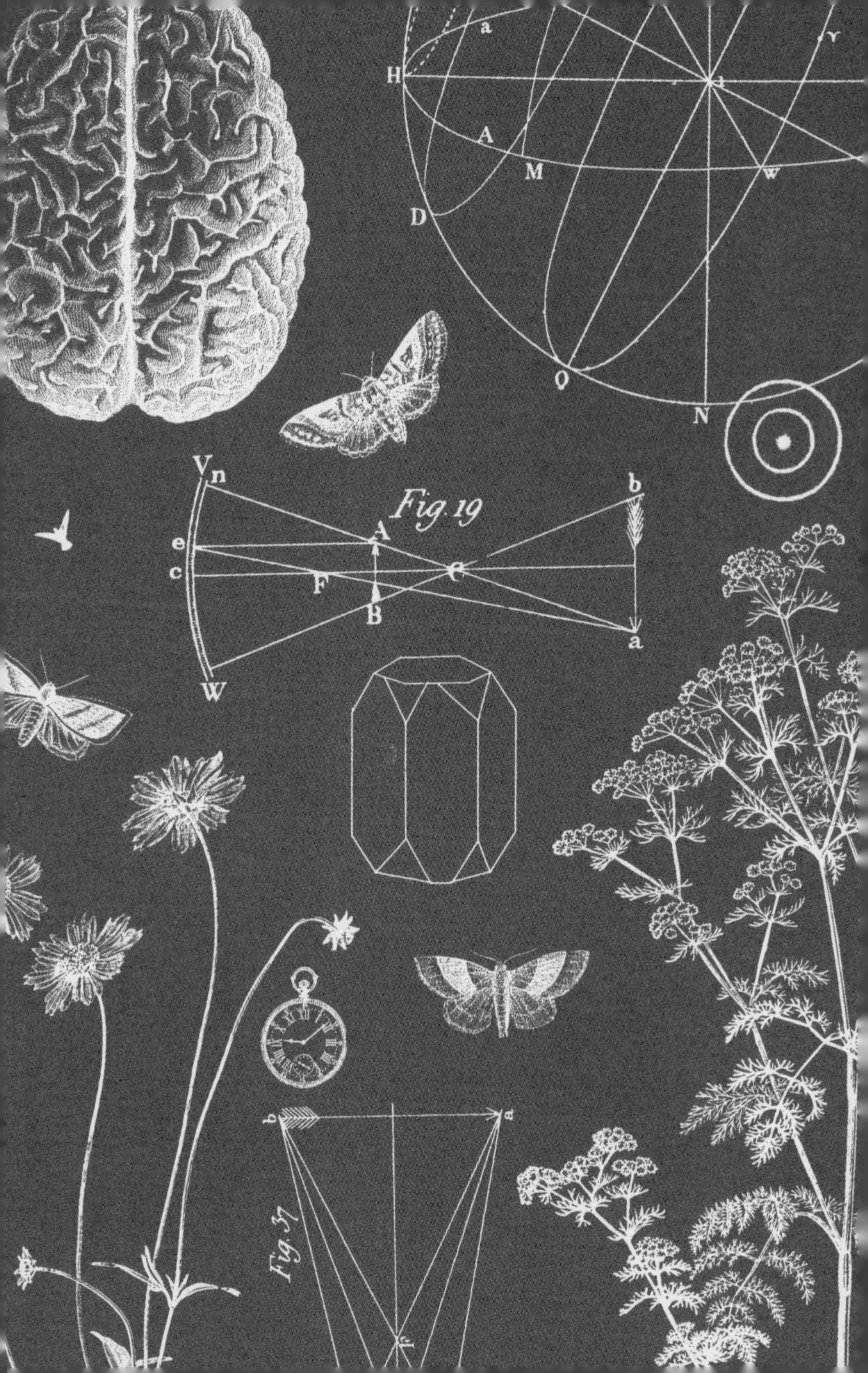

Fig. 19
Fig. 37

GET COMFORTABLE

The position you choose to sleep in could be a factor in your ability to sleep through the night. The most common sleep position – and the one recommended by many sleep experts – is foetal. If you choose to sleep this way you should favour the opposite side to the one of your dominance (in other words, if you're right-handed, choose your left side). Not all experts agree on this though, with many suggesting that sleeping on your back is better for your health, even though this is the least popular position to sleep in.

Establishing the best position for sleep ultimately comes down to comfort, and you can figure this out through trial and error. However, certain positions are more favourable if you suffer with specific health conditions that are affecting your sleep quality.

BACK AND NECK PAIN

Sleeping on your back allows your head, neck and spine to rest in a neutral position, which limits any excess pressure on those areas. Placing a pillow under the back of your knees can help to support the natural curve of your lower back and further lessen any stress on your spine. Make sure the pillow you rest your head on supports the natural curve of your neck and shoulders.

SNORING OR SLEEP APNOEA

Sleep apnoea is a condition that causes the airways to collapse during sleep, which leads to interrupted breathing. The condition can cause disrupted sleep and snoring. Avoid sleeping on your back as this encourages the base of your tongue and soft palate to collapse to the back wall of your throat, which causes snoring. Adopt a side position to help prevent this happening and to aid in opening up the airways. Placing a firm pillow between your knees can help reduce any stress on your hips and lower back.

REFLUX AND HEARTBURN

Many people struggle with reflux and heartburn, which is caused by stomach acid rising up into the oesophagus and throat. Pregnant women and people who are overweight are more prone to this condition. Lying on your back can make symptoms worse, but if this is how you sleep, elevate your head and shoulders to an incline using pillows. Sleeping on your side has been shown to help with reflux and heartburn, but the side you choose is important and is largely down to gravity. Given the position of your oesophagus, sleeping on your left-hand side means reflux is more easily drawn back towards the stomach.

KEEP MOVING

There is absolutely no doubt that keeping physically active benefits your health in a multitude of ways, both physically and mentally. Research shows that physical activity helps to reduce the risk of all-cause mortality by 30 per cent.

The current guidance is that we should keep active on a daily basis and exercise for 30 minutes each day, five days a week. Despite the benefits of this, a significant number of people still lead a sedentary lifestyle.

Exercise benefits sleep in a number of ways, not least in increasing the time spent in deep sleep (stages 3–4), which is the most physically restorative phase of your sleep cycle. While helping to improve the quality of your sleep, physical activity can also aid in lengthening the amount of time that you spend asleep. This is simply down to the fact that you are expending more energy across the day, which helps you to feel tired and ready to rest at bedtime.

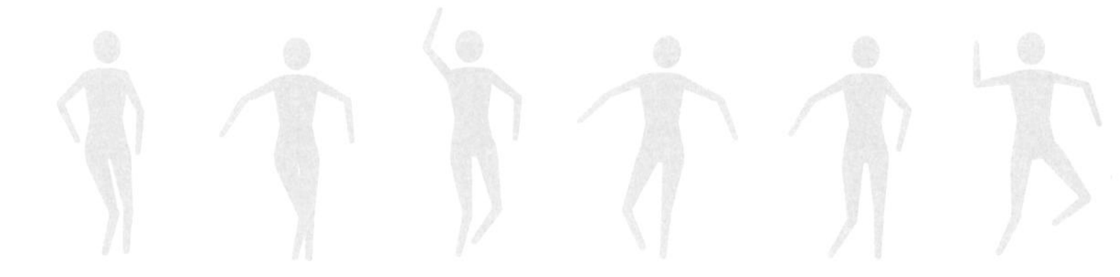

Research shows that this effect is more prevalent in people who exercise regularly as part of a consistent routine.

Exercise can also help to relieve stress and anxiety, which are common barriers to sleeping well. Mind-body activities such as yoga and stretching have been shown to help lower cortisol levels and reduce blood pressure, both of which can help to induce sleep.

But training too close to your bedtime may also impact on your sleep, especially if this involves high-intensity exercise. Heavy exercise significantly raises your heart rate, body temperature and levels of hormones such as adrenaline, all of which do not make for a perfect sleep cocktail. Training late often also means eating late, which may disturb sleep in some people. If you prefer to train late, try taking a cool shower after your workout to reduce your body temperature before bed, and make your evening meal the smallest of the day.

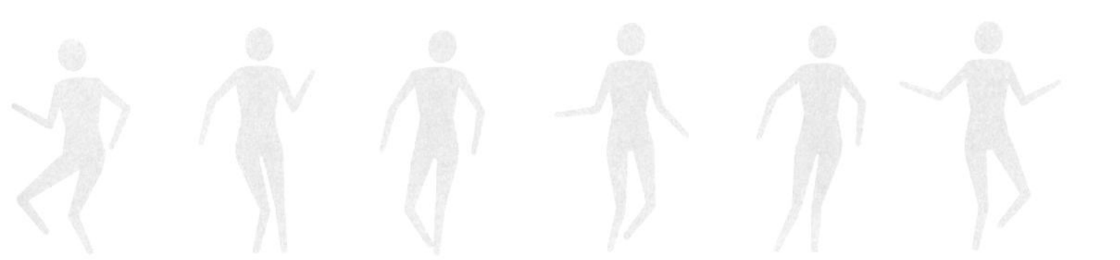

CHAPTER FIVE

environment

‘Have nothing in your house that you do not know to be useful, or believe to be beautiful.’

William Morris

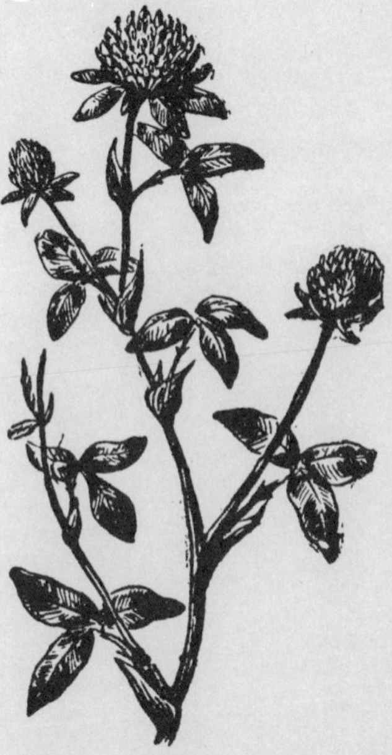

MESS CAUSES STRESS

People who struggle with sleep tend to become hypersensitive to anything that poses a 'threat' to drifting off.

Whether it's a ticking clock, cluttered shelves, disorganized wardrobes, heaps of dirty laundry, piles of work files or the standby lights on electrical equipment, it doesn't take much for any one of these things to become the focus of attention or even become something to obsess about when trying to fall asleep. Even little things that may seem insignificant during the day can become a source of anxiety, such as a peeling bit of wallpaper or a crack in the wall.

Mess causes stress, so it's important to declutter and create a quiet, calming sleep haven.

YOU'VE MADE YOUR BED

Now you need to lie in it! It really does make a difference to keep your linen fresh and your bed made up invitingly. Plump up those pillows, shake out your duvet so it's as fluffy as possible and create your own slumber paradise. White sheets can be particularly calming, but of course it's all down to personal preference. And don't forget about what's under the bed – it might be out of sight but it's certainly not out of mind. If you really need to use the space, invest in proper storage boxes to help tidy everything away.

DON'T SWEEP IT UNDER THE CARPET

If you don't need it in the bedroom, don't keep it in the bedroom. Remove anything that's not conducive to a restful, relaxing environment, save only your lamp and possibly a candle on your bedside table. And yes, the same rule applies to any digital devices cluttering your bedroom, such as TVs, laptops and phones, this is your space to sleep, so be strict with yourself.

SKELETONS IN THE CLOSET?

As a final step, work through your wardrobe, as knowing that everything in your room is decluttered, even if hidden behind a wardrobe door can help create a feeling of organization and promote a sense of clarity. Start by removing items of clothing that you don't wear any more by organizing your clothes by season and putting summer clothes away in winter, and vice versa.

Choosing your outfit the night before and laying it out ready for the morning is a simple way of helping to calm a busy mind by contributing to the sense of organization and eliminating any unnecessary stresses when you wake up.

Declutter and organize your room in small sections to lessen the load.

'Invest in your shoes and your sheets, because if you're not in one, you're in the other.'

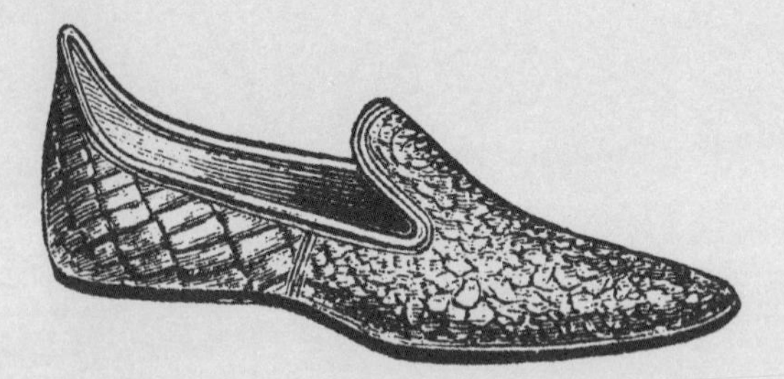

Yes, there really is truth in this old proverb! The surface you sleep on is incredibly important for promoting a good night's rest, so it pays to invest. Did you know that mattresses can deteriorate by as much as 70 per cent within ten years of use? As a benchmark, you should consider replacing your mattress every seven years.

Research shows that 20 per cent of people attribute back pain, neck pain, stiffness and other aches as the reason for not sleeping well, and these are just a few of the physical signs you experience that might hint that you need a new mattress.

If you find yourself suffering with an itchy or runny nose, sneezing or coughing (amongst other symptoms), this could also indicate a build-up of dust and allergens in your mattress – which is another reason to replace it.

Take your time, do the research and establish the right level of comfort to suit your needs. Should you choose pocket-sprung or memory foam? Each has its unique benefits, especially if you struggle with back pain. And if you're sharing your bed with a partner, it's best to go for the largest size possible, to give yourself room to spread out!

Fig.15
Fig.16
Fig.19
Fig.20
Fig.23
Fig.22
Fig.40
33
Fig.32
Fig.41
Fig.42
Fig.44
Fig.47
46
Fig.49
Fig.60

BEDDING DOWN FOR THE NIGHT

The type of bedding you choose to use can have an impact on your quality of sleep in many ways.

1. Make sure your bedding is as breathable as possible in order to keep your body temperature stable and in sync with your circadian rhythm.
2. Make sure you have the right weight of duvet, too, to keep your temperature steady. Many duvets come as two tog varieties clipped together, which is useful to get through the seasons and gives you two in one. Otherwise, as a rule of thumb, you should use a 13.5-tog duvet in winter, 9-tog in autumn and 4.5-tog in summer. If it's really hot you might prefer to use a separate sheet or the duvet cover on its own.
3. Whether you suffer with allergies or not, ideally your bedding should be hypoallergenic. Dust mites thrive on bedding as they dine out on your dead skin cells and they trigger allergic reactions in response to their droppings, which will affect your breathing through the night and wake you up as a consequence. Sleep tight, don't let the bed bugs bite!

7.b
7.f
12.h
12.d
Fig. 12.
7.i
a
4
4.c
k
12
l
f

CHAPTER SIX

diet

‘One cannot think well, love well, sleep well, if one has not dined well.’

Virginia Woolf

EAT RIGHT TO SLEEP TIGHT

Years of research have shown that diet is undoubtedly associated with good health. In the short term, the food we eat gives us the nutrients we require to fuel our day-to-day tasks.

Carbohydrates provide an immediate source of energy, while fats act as a way of storing it.

Proteins support the growth and repair of tissues throughout the body.

Vitamins and minerals are required in much smaller amounts and are essential to life as they help to support the many reactions that take place to keep our bodies in good working order.

The link between diet and sleep is not clear-cut, and nothing you eat is going to be a miracle cure, but when considered in tandem with good sleep hygiene and relaxation techniques, choosing what and when to eat will form a key part of your sleep ritual.

Certain food and drinks have been shown to help induce sleep by way of their nutrient content, as well as those that have the potential to keep you awake.

Anxiety, depression and stress can also affect our appetite and nutrient intake in several ways, as well as put greater demand on the body for particular food types, some of which have been linked to sleep. Nutrient insufficiencies can be caused by a lack of appetite as well as overeating.

Use the findings from your sleep diary and the information in the following chapters to help you to think about what and when to eat in order to promote the best night's sleep.

Eating rubbish food isn't a reward, it's a punishment.

The
WORCESTER
BRAND
SALT
NASH, WRITON & Co.
NEW 120, WARREN ST. YORK
BITTERS
18b
Liqueur
Pères Chartreux
Bas de Carré (2)
Collet (3)
Tête (2)
Carré (1)
Épaule (2)
Grosse Poitrine (2)
Crosse (3)
BORDEN'S
Peerless
"Brand"
EVAPORATED
CREAM
UNSWEETENED

ALCOHOL

The most common self-medicated sedative is alcohol, but its effect on sleep is tricky and something of a double-edged sword. It's true that a little alcohol may help you to relax, but even in small amounts it can cause fragmented sleep and could be considered an undercover sleep marauder.

While the seemingly harmless nightcap might be relaxing at first, it can have rebound effects, causing you to wake up during the night due to dehydration or the need to visit the bathroom, and in some cases it can also contribute to heartburn.

Alcohol can impair the restorative part of the sleep cycle, REM, as well as interfering with the flow of calcium into nerve cells, affecting the region of the brain that controls sleep function.

If you don't wish to cut out alcohol completely, instead keep your intake before bed to a minimum. Enjoy a drink with your evening meal a few hours before bedtime to ensure that it has the least effect on sleep.

SPICY FOODS

On your quest for a good night's sleep, your favourite curry could hinder all good efforts. Spicy foods can cause or exacerbate heartburn in people prone to indigestion. No matter how much you enjoy these foods you need to be realistic when deciding what to eat, and if you're more prone to indigestion they are better avoided.

CAFFEINE

Avoid caffeine. That's obvious, right? We're all aware of the stimulating effect caffeine offers, which is why it's such a popular drink in the morning to pep us up for the day ahead.

Caffeine is a stimulant that blocks the chemicals in the brain that make you sleepy. It is not only found in coffee but also tea, soft drinks and chocolate. Caffeine can remain in the body for 3–5 hours, but in some people the effects can still be seen twelve hours after consumption. Not everyone has the same reaction to caffeine, though, and this is down to a gene called CYP1A2 that controls the enzyme determining how fast you metabolise it. If you're lucky enough to have the fast variant of the gene, you metabolise caffeine four times faster than those with the slow variant.

Regardless of whether you're sensitive to caffeine or not, in the interest of sleep it's worth avoiding it in the 6–8 hours before going to bed while you establish your sleep ritual. By completing your sleep diaries you'll be able to establish the effect of caffeine on your sleep pattern.

Instead, save your caffeine fix for the morning as a kick-start to your day. Beyond midday, switch to caffeine-free hot beverages such as rooibos tea, or for something with more of an energising kick go for ginger or ginseng. Herbs such as lemon balm, valerian and chamomile have long been used as teas for their relaxing effects and are a good option before bedtime.

TYRAMINE

We have all heard the old wives' tale that eating cheese before bed gives you nightmares. The thing with old wives' tales is that there's often a little truth behind them. Research has shown that there might actually be something behind this one. Cheese, along with other foods such as bacon, ham, aubergines, pepperoni, avocado, nuts, soy sauce and red wine, contain an amino acid called tyramine. This amino acid is a common trigger for people who suffer with migraines but it

GOODALL BACKHOUSE & Co
LEEDS
SEAL BRAND
JAVA
COFFEE
BOSTON
AMERICAN
FOOD Co
FRENCH
SOUPS
NEW YORK
TURTLE
PRESERVERS
BLUE LABEL
TOMATO
KETCHUP
FRENCH
DRESSING
FOR
LADIES' AND
CHILDRENS
BOOTS & SHOES
BOSTON, MASS
ENGLAND
MEAT
PURITY

might also inhibit sleep as it causes the release of a hormone called norepinephrine that can stimulate the brain.

These foods will not definitively prevent you from sleeping, but if you do want to investigate the effect of diet on sleep then you can experiment with them to see if they appear to have any impact by cutting them out and reintroducing again. Only cut one out at a time to help establish their potential impact on sleep.

SUGAR

Food and nutrition surveys have shown that adults eat twice the recommended amount of six teaspoons of sugar per day. White sugar is referred to as a 'free' sugar, which is found in all sweeteners (including honey, agave and fruit syrups). The main contributors to increased sugar levels in the diet are soft drinks and table sugar (added to foods and drinks) followed by confectionery, puddings and desserts.

Eating lots of sugar during the day can impact on your quality of sleep during the night and pull you out of a deep sleep. One study showed that high sugar consumption led to less deep sleep and more arousals. Sugar also reduces the activity of orexin cells, which stimulate parts of the brain that produce dopamine and norepinephrine, which are two hormones that keep us aroused and physically mobile.

Researchers at Cambridge University have found that orexin cells are sensitive to glucose levels, which means that when you have high levels of glucose in the blood their activity is reduced and you can feel sleepy. Energy slumps during the day are naturally part of your circadian rhythm, but enhancing them with too much sugar can encourage napping during the day, which can impact on sleep later in the evening.

Interestingly, the same research also found that amino acids (protein) can not only stimulate orexin cells but also prevent glucose from inhibiting their function. This means that opting for a lunch or any snacks rich in protein can help to prevent debilitating mid-afternoon slumps in energy, which can feel more pronounced in those who suffer from sleep deprivation.

de Milan.

Fig. 477. — Chou fleur.

Choux de Bruxelles.

icorée frisée.

Fig. 481. — Barbe de Capucin.

Mâche ou Dou

Asperges.

Griffes ou rhizom

Céleri rave.

Fig. 490. — Persil.

TRYPTOPHAN

Seeds (sunflower, pumpkin, hemp, chia), nuts (cashew, almond, hazelnut), soya foods (soya beans, soya milk, tofu), bananas, cheese, meat (beef, lamb, pork), poultry (turkey, chicken), oily fish (salmon, tuna, trout), oats, beans, lentils and eggs.

Tryptophan is an essential amino acid that must be obtained from the diet as the body can't make it. It is required to make melatonin in the brain, which is the hormone that helps to make you feel drowsy and ready for sleep. This amino acid has a lot of competition to make it across the blood brain barrier, but including foods rich in carbohydrates, such as pasta, rice and potatoes, can increase its uptake. These foods elevate the hormone insulin, which helps the uptake of tryptophan in several ways, such as reducing the levels of other amino acids that compete with it for transport carriers to the brain. When planning what to eat, a combination of tryptophan-rich foods teamed with carbohydrates may be the perfect option for an evening meal.

VITAMIN B6

Pulses and lentils (chickpeas, brown lentils), liver, oily fish (salmon, tuna, trout), meat (beef, lamb, pork), poultry (turkey, chicken), bananas, soya foods (soya beans, soya milk, tofu).

One of the roles of vitamin B6 in the body is for the production of melatonin, which is the hormone that controls the sleep-wake cycle. On the whole, most of us get enough vitamin B6 as it's available in many foods, but it's also easily depleted as a result of stress or excessive alcohol intake. While planning your sleep diet, be sure to include plenty of foods rich in vitamin B6 to keep levels topped up.

MAGNESIUM

Dark green leafy vegetables (kale, spring greens, spinach), seeds (sunflower, pumpkin, hemp, chia), beans and pulses (red kidney beans, chickpeas, soya beans), lentils, oily fish (salmon, tuna, trout), wholegrains and pseudo grains (quinoa, brown rice, bulgur, whole wheat pasta and bread), nuts (cashew, Brazil, walnuts), cocoa (raw cacao, dark chocolate) and avocado.

This mineral is one of the most abundant in the body and has many functions that include proper bone, brain, heart and muscle function. Magnesium activates the parasympathetic nervous system, which is responsible for relaxation. This mineral binds to gamma-aminobutyric (GABA) receptors responsible for quieting nerve activity and by doing so may help prepare your body for sleep. Magnesium also regulates melatonin, which guides sleep-wake cycles in the body. Include plenty of magnesium-rich foods in your diet from the list above.

CALCIUM

Dairy foods (milk, yoghurt, cheese), tofu, fortified milk alternatives including soya and nut milks, dark green leafy vegetables (kale, spring greens, spinach), beans and pulses (red kidney beans, chickpeas, soya beans), dried fruits, dried spices, canned fish (salmon, pilchards, anchovies), squash (butternut, acorn) and shellfish (crab, lobster, prawns).

Calcium is a mineral required to convert tryptophan to melatonin, and research published in the *European Neurology* journal has found that disturbances in sleep, especially during REM, may be related to low levels of calcium. Be sure to include a good supply of calcium from the foods listed above in your diet.

B
C
G
N
M
E
C
I
C
C
IV

BREAK BAD HABITS

OVERWEIGHT

Not only is being overweight bad for your health, it can also affect your mental wellbeing and self-confidence, which can impact on sleep quality. Sleep apnoea is a disorder often associated with being overweight, which interferes with your breathing patterns and interrupts your sleep.

On the flip side, research has suggested that a lack of sleep may be a factor in weight gain. A lack of sleep can cause tiredness, which provides a barrier to exercise. Being awake longer means more time and opportunity to eat. A lack of sleep also disrupts that balance of hormones that control appetite, meaning sleep-deprived people may be hungrier than those who get enough rest each night.

If you're overweight, then make trying to lose weight a priority if you have difficulty sleeping.

HEARTBURN AND INDIGESTION

Indigestion is a common complaint, but for some people it can be something they have to deal with on a daily basis. The condition is normally the result of inflammation in the stomach caused by excess stomach acid and it can easily interrupt sleep or make it difficult to nod off. One of the simplest ways to tackle indigestion is to eat smaller meals throughout the day; in order to aid sleep you should look to eat 2–3 hours before bed. Citrus fruits, coffee and tea all promote high acid production so are best avoided.

Following a low-fat diet and eating meals made up of a good balance of starchy foods, protein (which stimulates the gall bladder to produce more bile to aid digestion) and vegetables is a good place to start. Very rich foods high in fat can cause problems as they take a long time to digest. Including oily fish in your diet may help, as the omega-3 fatty acids found in these fish can help to reduce inflammation and encourage better digestion.

Taking your time to eat and chewing your food properly can help to stimulate enzymes that aid digestion. You should also avoid fizzy drinks, mint and chocolate as they can relax the gut wall and encourage reflux. Raw vegetables can also contribute to indigestion, so try to avoid them and opt for cooked vegetables as a preference as they are easier to digest.

LIQUID INTAKE

Drinking too much before bedtime can cause you to wake up to use the bathroom, so limit your intake of liquid a couple of hours before you go to bed.

HERBS AND SUPPLEMENTS

Supplements are a useful way to bridge any gaps in the diet that may be impacting on your ability to sleep, and in some cases certain herbs or micronutrients (vitamins and minerals) may help to induce sleep.

If you're looking for a natural comparative to a sleeping pill, think again. Believe me, I have tried most supplements promising a good night's sleep and on the whole I still lie in bed counting sheep, waiting for something to 'kick in'. I'm very pro-supplements, though, and while my first choice of nutrition is always going to be food, if used in the right way supplements may have a beneficial effect, but it is often a subtle one and mostly beneficial if your diet is lacking in a certain nutrient. If you're considering taking a supplement then do so for a few months then stop for a month to help establish whether their impact is beneficial.

MAGNESIUM

Magnesium is involved in the regulation of melatonin and a study published in the *Journal of Review of Medical Sciences* showed how taking a magnesium supplement helped to improve levels of melatonin, sleep time and sleep efficiency (reduction in waking up during the night).

Magnesium supplements are often used as a way of encouraging better sleep, relieving aching muscles and balancing mood, especially for women with PMS and during the menopause.

The National Diet and Nutrition Survey carried out in the UK has highlighted an insufficient intake of magnesium amongst 13 per cent of adults, and teenage girls are a particular concern, with 50 per cent shown to have very low intakes of this mineral from their diet. On top of this, research has shown that we only manage to absorb about 50 per cent of the magnesium from the foods we eat. Stress also affects the body's requirements for magnesium and low levels have been shown to upset sleep as well as impacting on tiredness and fatigue.

Try taking a magnesium supplement (375mg) before you go to bed. Magnesium can also be absorbed through the skin, so using magnesium salts in the bath before bed may also help.

5-HTP

5-hydroxytryptophan (5-HTP) is a unique amino acid found naturally in the West African medicinal plant *Griffonia simplicifolia*. As mentioned, tryptophan is involved in the production of the brain chemical melatonin that helps to regulate sleep.

5-HTP appears to improve sleep architecture by extending the amount of time spent in REM, so in theory helping you to wake up more refreshed. Research has shown how 5-HTP may be particularly useful for sleep disturbances associated with fibromyalgia (a condition that is characterized by muscle and bone pain as well as general weakness), as it may help to reduce pain perception.

As with all supplements, you need to give 5-HTP a while before you notice any effects. Start with 100mg every night, building up to a maximum of 300mg, and use for three months before you review the impact of this supplement.

VALERIAN

This traditional herbal remedy is associated with the treatment of stress, but may also help to aid sleep. The sedative effect of valerian is due to the inhibition of enzymes that break down an inhibitory brain chemical called GABA, gamma-aminobutyric acid. As GABA levels rise it dampens down the over-stimulation that can cause anxiety-driven thoughts that stop you sleeping. Valerian has been shown to be particularly useful for women during the menopause.

Valerian is available in capsules, tinctures and teas. If you're using the tinctures or teas, it does have quite a pungent smell that takes a little getting used to.

PLANNING IS EVERYTHING!

You should aim to eat three nutritious meals each day. When you are planning what to eat, keep the foods mentioned in the previous section in mind – both those that harm and heal.

Also take into consideration any drinks such as alcohol or coffee that you may have highlighted in your sleep diary.

Make sure you have healthy snack ideas to hand if you'll need to skip a meal or go long periods in between eating. Both these instances lessen the opportunity for you to glean essential nutrients from the diet and can leave you lacking in the energy required to get you through the day. Remember though, if you are eating three nourishing meals each day then there's no need for a snack unless your lifestyle requires more energy. Unhealthy snacking, especially on sugar-laden foods and drinks can contribute to weight gain and also heighten blood sugar imbalances which can impact on energy levels.

When planning your diet, also consider when you will eat your last meal of the day (hopefully one full of sleep-promoting nutrients) in order to aid a good night's sleep. You may, for example, choose to eat earlier or eat a smaller meal if you are busy late in the day. Also, remember those foods that may impact on indigestion and consider these when planning what to eat before bed time.

Planning your diet in advance is a good way to insure you have everything in store to create sleep-friendly meals and try to stick to your plan as closely as possible, but don't be too hard on yourself if life gets in the way – it will happen!

Here are a few examples of how to incorporate more sleep-friendly foods into your daily diet.

BREAKFAST

- **Yoghurt topped with nuts, seeds and dried fruits** *(rich in calcium, magnesium and B6)*
- **Scrambled egg with spinach** *(rich in magnesium, B6 and tryptophan)*
- **Breakfast smoothie made with fortified soya milk, berries and oats** *(rich in tryptophan, magnesium, calcium and B6)*
- **Avocado on toasted rye bread** *(rich in magnesium, B6 and tryptophan)*

- **Greek salad with wholemeal pitta bread**
  (rich in calcium, tryptophan, B6 and carbohydrates)
- **Chicken and salad wrap**
  (rich in B6, tryptophan and carbohydrates)
- **Mixed bean or wholegrain salad with protein such as chicken, canned tuna, halloumi cheese or grilled tofu**
  (rich in tryptophan, B6, calcium and carbohydrates)
- **Frittata made with vegetables such as sweet potato or peppers and the option of feta or goat's cheese**
  (rich in calcium, tryptophan, B6 and calcium)

EVENING MEALS

- **Tofu stir-fry with buckwheat noodles, quinoa or brown rice** *(rich in magnesium, calcium, B6, tryptophan and carbohydrates)*
- **Wholemeal spaghetti with beef Bolognese sauce** *(rich in B6, tryptophan and carbohydrates)*
- **Mushroom risotto made with rice or spelt** *(rich in B6, magnesium and carbohydrates)*
- **Roast chicken with new potatoes and multicoloured vegetables** *(rich in magnesium, B6, tryptophan and carbohydrates)*

SNACKS

- **Cashew nuts** *(rich in magnesium and B6)*
- **Plain yoghurt with berries** *(rich in calcium, B6 and tryptophan)*
- **Wafer-thin turkey slices on rye crispbreads** *(rich in B6, tryptophan and magnesium)*
- **Cheese and oatcakes** *(rich in B6, calcium, tryptophan and carbohydrates)*

CASHEW NUT MILK WITH RAW CACAO

Serves 2

This homemade chocolate nut milk is loaded with magnesium, which helps the muscles to relax and facilitates the production of melatonin in the brain.

150g raw cashew nuts

800ml water

3 level tbsp raw cacao powder

1 tbsp honey (or maple syrup if you're vegan)

1 vanilla pod, seeds scraped out (optional)

Pinch of sea salt

1. Soak the cashews in a bowl of water for 3 hours (or you could soak them overnight the day before).
2. Drain the cashews and add them to the blender with the fresh 800ml water.
3. Add the remaining ingredients and blend for a minute on high or until completely smooth.
4. Store for up to three days in a clean glass bottle in the fridge or serve immediately (best served really chilled).

WARM ALMOND CHAI MILK

Serves 2

This homemade milk is full of magnesium to encourage the muscles in your body to relax and compounds that help to reduce inflammation, which is one of the negative aspects of not sleeping for a prolonged period of time.

500ml almond milk

½ tsp ground turmeric

¼ tsp ground cinnamon

¼ tsp ground cardamom pods

2 tsp honey

Small pinch of sea salt

1. Place all of the ingredients in a small saucepan and set over a medium heat.
2. Warm the milk slowly and be careful not to let it boil.
3. Once warmed, serve in small mugs.

LEMON BALM, LAVENDER AND LICORICE ROOT TEA

Serves 2

Lemon balm and lavender are two herbs that have long been known for their relaxing properties, which may help to promote sleepiness before bedtime. Licorice has a sweet aniseed taste and can help to reduce the need for adding sugar to drinks.

1 tbsp dried lemon balm
1 tbsp dried peppermint leaves
1 tsp fennel seeds
1 tsp dried rose petals
1 tsp dried lavender flowers
2 slices dried liquorice root
Honey, to taste

1. Add all the ingredients to a teapot and then fill with boiling water.
2. Leave to brew for 5 minutes then pour through a tea strainer (unless you have a diffusing teapot).

CHAPTER SEVEN

mindfulness

‘Sleep is the best meditation.’

Dalai Lama

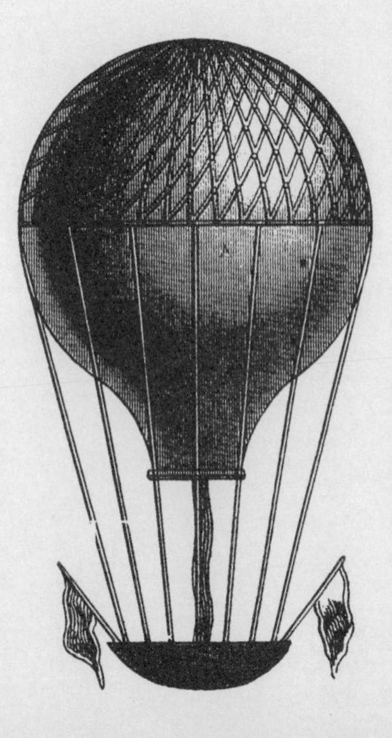

LOSING SLEEP OVER IT? Read through these three key indicators to help you decide if stress is at the root of your own sleep deprivation.

YOU CAN'T TURN OFF YOUR BUSY MIND

You keep going over and over your stresses, worries and frustrations, contemplating them from various angles. It's almost like they're playing on a continuous loop that you can't shut off, which interferes with your ability to slip into slumber.

YOU ARE TENSING YOUR MUSCLES

You're experiencing muscle tension and pain, or stress-related aches such as neck and shoulder pain or headaches, and it can be difficult for you to fall asleep or stay asleep. Complicating matters, poor sleep can set the stage for you to experience even more tension headaches and increased pain sensitivity the next day.

Your heart rate is revved up and variable, associated with increased levels of cortisol (a stress hormone), greater physical tension and increased autonomic arousal, which affects your ability to fall asleep or sleep well.

If you can identify with one or more of these reasons, try experimenting with various relaxation techniques to soften the mind and reduce stress.

Make relaxation your goal, not sleep.

RELAX EVERY MUSCLE

Relaxation techniques can help you achieve physical and mental relaxation by reducing tension and interrupting the thought processes that can disrupt sleep.

Try this twenty-minute relaxation technique at any time of day, but particularly before bed, to slowly release tension one muscle group at a time.

Sit comfortably or lie down in a quiet place, hands resting down by your side.

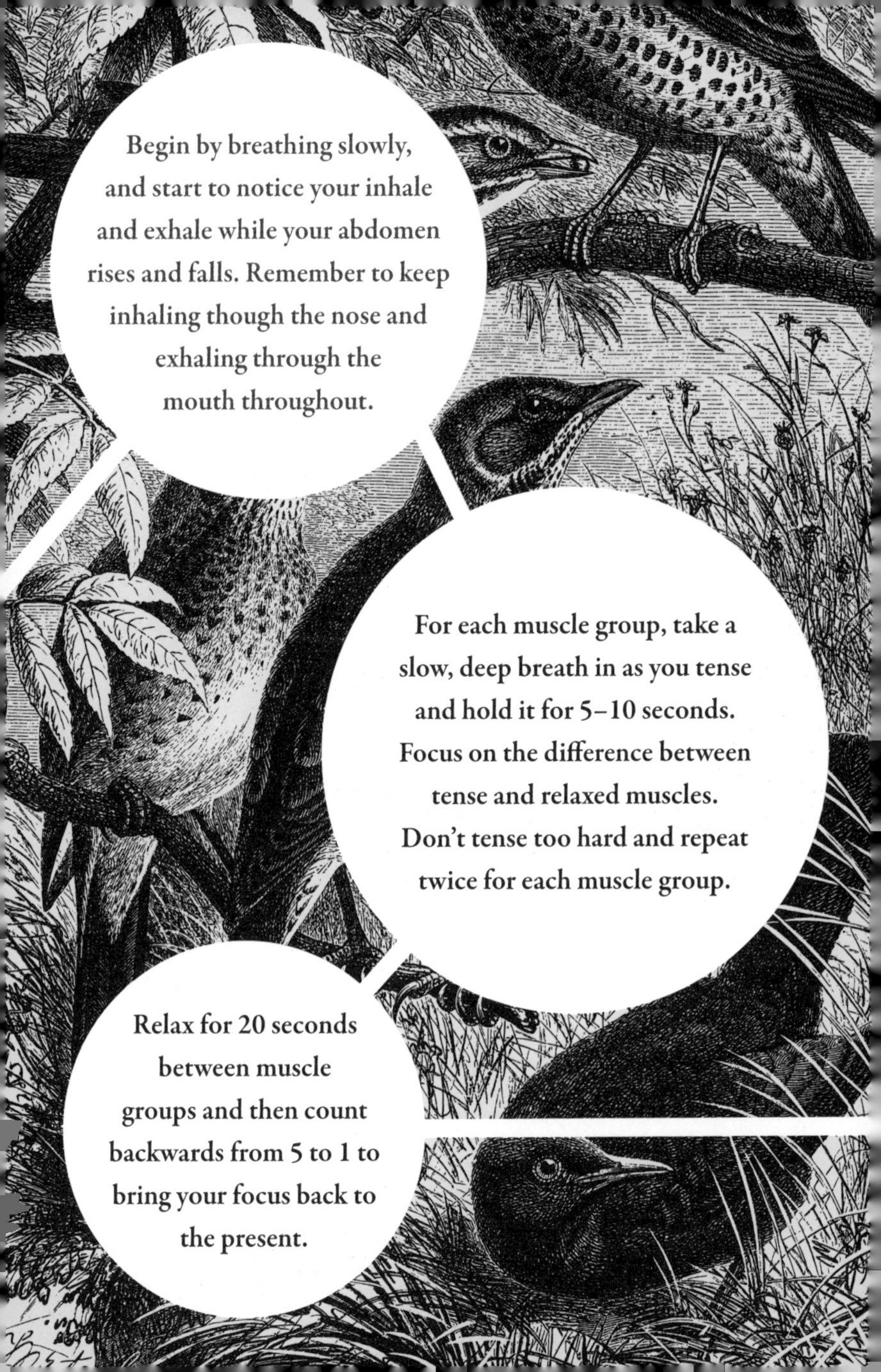

Begin by breathing slowly, and start to notice your inhale and exhale while your abdomen rises and falls. Remember to keep inhaling though the nose and exhaling through the mouth throughout.

For each muscle group, take a slow, deep breath in as you tense and hold it for 5–10 seconds. Focus on the difference between tense and relaxed muscles. Don't tense too hard and repeat twice for each muscle group.

Relax for 20 seconds between muscle groups and then count backwards from 5 to 1 to bring your focus back to the present.

Foot

Curl your toes downwards.

Lower leg and foot

Tighten your calf muscle by pulling your toes towards you.

Entire leg

Squeeze your thigh muscles and pull your toes towards you, then repeat on the other side of your body.

Hand

Clench your fist.

Arm

Tighten your bicep by drawing your forearm up towards your shoulder to 'make a muscle', while clenching your fist at the same time. Then repeat on the other side of your body.

Buttocks

Tighten them by pulling your buttocks together.

Stomach

Suck your stomach in.

Chest

Take a deep breath to tighten the chest.

Neck and shoulders

Raise your shoulders up to touch your ears.

Mouth

Open your mouth wide enough to stretch the hinges of your jaw.

Eyes

Clench your eyelids tightly shut.

Forehead

Raise your eyebrows as far as you can.

THE STUFF OF DREAMS

Visualizing your own personal nirvana is a valuable exercise that can help you to unwind, relieve stress and fall asleep. Unlike meditation, visualization is more active as your breath and mind are guided in a specific direction to achieve a desired result. This technique helps to direct your attention away from the thoughts that may be causing stress and anxiety by associating sensations of relaxation with peaceful images in the mind.

To start, find a quiet place free from distraction and make yourself comfortable. Take a few deep breaths to centre yourself and then close your eyes. Visualize yourself in a location where everything is ideal to you and imagine yourself becoming calm and relaxed or feeling happy and care-free. Focus on every sensory element of your scene to create a vivid image, and take time to explore these elements in your mind until you feel relaxed. Hold on to this image and assure yourself that you can return to this place again to help you to relax before opening your eyes.

Fig. 14
Fig. 15
Fig. 7.
Fig. 12.

Guided visualization such as the examples on the next few pages can help to lead you through the technique to start with. As you continue to practice you will be able to go more deeply and quickly. The examples here are brief, and you can find longer scripts online. There are also many apps available to help talk you through guided visualization.

Remember to breathe deeply through these exercises. Focus on breathing through your diaphragm by placing one hand on your belly to feel the rise on the inhale and the dip on the exhale.

Place your other hand on your upper chest to make sure it's still. Most of us breathe through our chest, which can maintain the stress response.

Deep breathing triggers the para-sympathetic nervous system to calm and put the body into a stable state.

Choose your favourite place, one where you feel calm and safe. It could be a garden, a waterfall, a room, or anywhere that takes your fancy. Now, close your eyes, visualize being there and relax into breathing deeply.

Walk around slowly and notice the colours and textures around you. What do you see? What do you feel? What do you hear? What do you smell? Take your time while you walk around. Spend some time exploring each of your senses.

Notice how peaceful and relaxed you feel, and remember this feeling. Say to yourself: 'I am relaxed, my body feels warm and heavy, I am safe here'. Enjoy the feeling of deep relaxation.

When you are ready, gently open your eyes and come back to the present moment.

A DAY AT THE BEACH

Imagine it's a beautiful sunny day and you're walking on a beach.

Now, close your eyes, visualize being there and relax into breathing deeply.

The sky is blue, the water is crystal clear, you hear the sound of gentle rolling waves as the light breeze caresses your skin.

The white sand feels warm on your bare feet and trickles between your toes.

You are wearing light flowing clothes and breathing deeply, inhaling the smell of fresh ocean air.

A sense of freedom washes over your body, you lie down and let your body sink into the warm soft sand.

Let go of any tension and soften your eyes while continuing to breath in sync with the lapping waves.

You sink deeper and deeper into relaxation

Say to yourself: I am relaxed, my body feels warm and heavy,

I am safe here.

DON'T FORGET TO BREATHE

There are many helpful breathing exercises you can do for relaxation before bedtime, such as this simple 4–7–8 technique developed by Dr Weil. The most important part of this process is holding your breath, as this allows oxygen to fill your lungs and circulate around the body, producing a relaxing effect throughout the whole body.

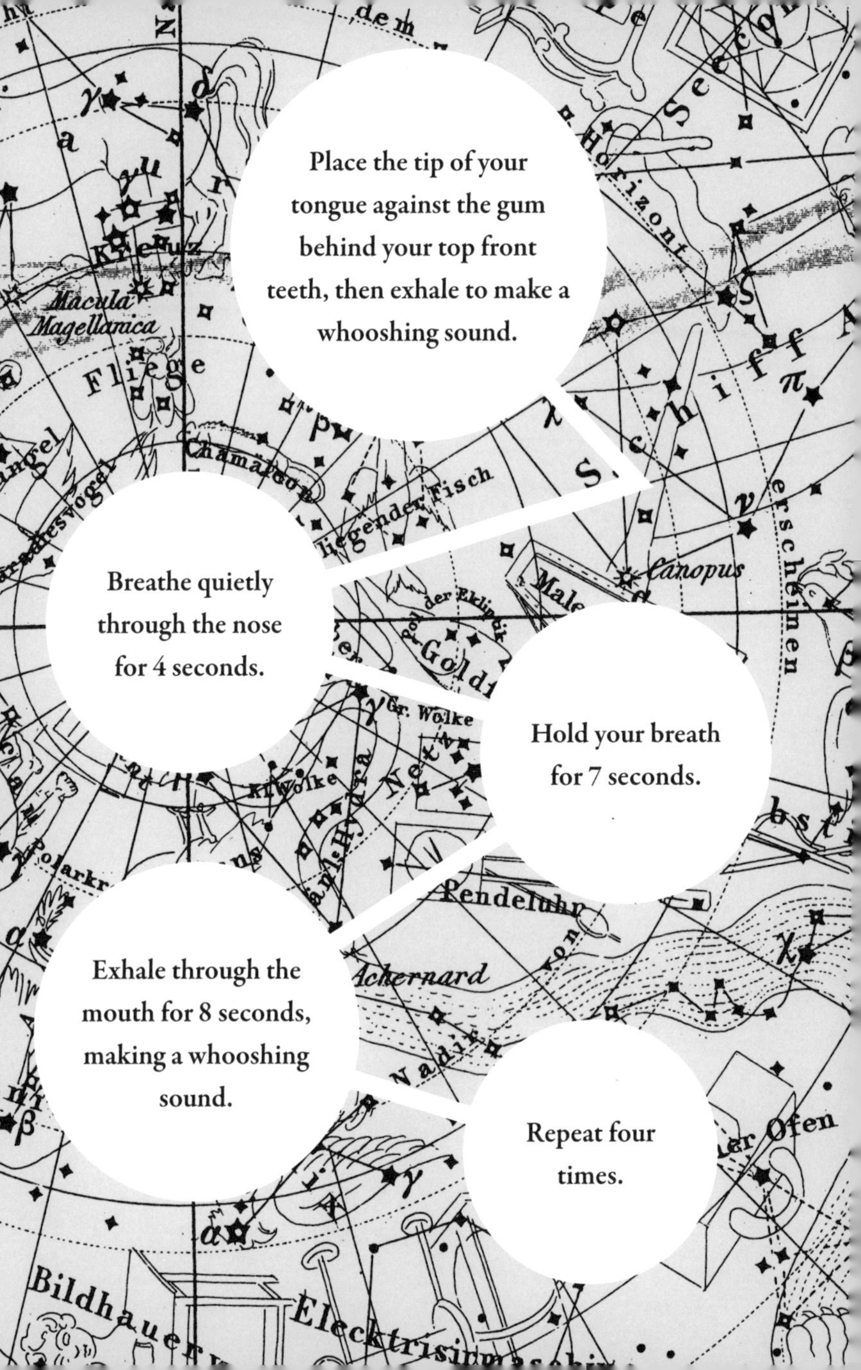
Place the tip of your tongue against the gum behind your top front teeth, then exhale to make a whooshing sound.
Breathe quietly through the nose for 4 seconds.
Hold your breath for 7 seconds.
Exhale through the mouth for 8 seconds, making a whooshing sound.
Repeat four times.

CHAPTER EIGHT
ritual

Whether you have read this little book from cover to cover or just dipped in and out, I hope it has helped you to understand more about why and how we sleep.

Now that you have more insight into your own personal sleep landscape, you can adapt your behaviour, environment and diet to develop your unique sleep ritual.

Don't panic if you still struggle to fall asleep.

Remember, resting
and relaxation (while
not a replacement for
sleep) still help to
rejuvenate your body.
The true secret to the
art of sleeping is finding
a ritual that works
best for you.

Fig. 34
F
a
b
c
d
e
f
g

ACKNOWLEDGEMENTS

I am hugely grateful to my wonderful agent Dorie Simmonds for driving me crazy to get this book off the ground! Her tenacity and confidence in me is what led to the publication of this little book. Thank you to all the team at HQ for their hard work and creative input that has made this book visually stunning and in particular Steve Wells who did the design and layout that brought this book to life. Thanks also to Charlotte Mursell for her guidance and direction whilst keeping everything on track. Finally, thank you to all the experts and non-sleepers who have shared their insight and knowledge with me on all things sleepy.

ABOUT ROB HOBSON

Rob Hobson is a registered nutritionist with a degree in nutrition, master's degree in public health nutrition and over 12 years of experience. Rob's work spans both public health, working within the NHS and alongside charity organisations, and industry, working with many of the UK's leading health and wellness brands. His writing is regularly featured in the UK health media, having written hundreds of articles for publications including the *Daily Mail, Daily Express* and *Women's Health* as well as being a regular voice on both radio and TV. Rob is co-author of the hugely successful book, *The Detox Kitchen Bible*, which has been translated into several different languages. Rob is passionate about health and his simple, realistic approach to living well is infectious and the result of both expert knowledge and personal experience. Rob shares this passion with private clients across the globe working as a nutritionist, healthy eating chef and sleep coach.